Marine Biology: A Very Short Introduction

T0021771

VERY SHORT INTRODUCTIONS are for anyone wanting a stimulating and accessible way into a new subject. They are written by experts, and have been translated into more than 45 different languages.

The series began in 1995, and now covers a wide variety of topics in every discipline. The VSI library currently contains over 600 volumes—a Very Short Introduction to everything from Psychology and Philosophy of Science to American History and Relativity—and continues to grow in every subject area.

## Very Short Introductions available now:

## Available soon:

For more information visit our website

www.oup.com/vsi/

Philip V. Mladenov

# MARINE BIOLOGY

## A Very Short Introduction

### SECOND EDITION

OXFORD
UNIVERSITY PRESS

# OXFORD

UNIVERSITY PRESS

Great Clarendon Street, Oxford, OX2 6DP,
United Kingdom

Oxford University Press is a department of the University of Oxford.
It furthers the University's objective of excellence in research, scholarship,
and education by publishing worldwide. Oxford is a registered trade mark of
Oxford University Press in the UK and in certain other countries

© Philip V. Mladenov 2020

The moral rights of the author have been asserted

First edition published in 2013
Second edition published in 2020

Impression: 5

Published in the United States of America by Oxford University Press
198 Madison Avenue, New York, NY 10016, United States of America

British Library Cataloguing in Publication Data
Data available

Library of Congress Control Number: 2019952749

ISBN 978-0-19-884171-5

Printed and bound by
CPI Group (UK) Ltd, Croydon, CR0 4YY

Links to third party websites are provided by Oxford in good faith and
for information only. Oxford disclaims any responsibility for the materials
contained in any third party website referenced in this work.

*For Taman, Maeve and Jem—may they enjoy the oceans as much as I have*

# Contents

# Acknowledgements

I am grateful to Jenny Nugee and the team at Oxford University Press for their advice and encouragement throughout this project and to Roger Harris for his careful reading and astute comments on the manuscript.

# List of illustrations

# List of abbreviations

| | |
|---|---|
| **AMOC** | Atlantic Meridional Overturning Circulation |
| **CCAMLR** | Commission for the Conservation of Antarctic Marine Living Resources |
| **DOM** | dissolved organic matter |
| **DSL** | deep scattering layer |
| **ENSO** | El Niño Southern Oscillation |
| **FAO** | Food and Agriculture Organization of the United Nations |
| **HAB** | harmful algal bloom |
| **HNLC** | high nutrient–low chlorophyll |
| **ICES** | International Council for the Exploration of the Sea |
| **IUCN** | International Union for Conservation of Nature |
| **IUU** | illegal, unreported, and unregulated |
| **MBARI** | Monterey Bay Aquarium Research Institute |
| **MPA** | marine protected area |
| **MSC** | Marine Stewardship Council |
| **NASA** | National Aeronautics and Space Administration |
| **NGO** | non-governmental organization |
| **POM** | particulate organic matter |

| | |
|---|---|
| **ppm** | parts per million |
| **UNCLOS** | United Nations Convention on the Law of the Sea |
| **UNEP** | United Nations Environment Programme |
| **USGS** | United States Geological Survey |
| **WCMC** | World Conservation Monitoring Centre |
| **WoRDSS** | World Register of Deep-Sea Species |
| **WWF** | World Wildlife Fund |

# Introduction

The oceans are our planet's most distinctive and imposing natural habitat. They cover 71 per cent of its surface; create a vast globally connected fluid living space; support a remarkably diverse and exquisitely adapted array of life forms from microscopic viruses, bacteria, and plankton to the largest existing animals; and possess many of Earth's most significant, intriguing, and inaccessible ecosystems. For these reasons alone, marine biology—the study of how oceanic organisms live and interact with each other and their environment—is an inherently significant and fascinating subject that has fostered our wonder, curiosity, and exploratory urges for generations.

However, now that Earth has entered the Anthropocene, a new geological epoch in which humans are significantly altering the global environment, the oceans are undergoing rapid and profound changes. The study of marine biology is thus taking on added importance and urgency as people struggle to understand and manage these changes. We now fully appreciate that functional ocean ecosystems provide services that are essential for human survival and well-being—they produce half of the oxygen we breathe; stabilize our climate; sustain ecosystems that protect our coasts; provide us with abundant healthy food; host diverse organisms that provide us with natural products for medicine and biotechnology; and support many forms of recreation and tourism.

The value of healthy, intact ocean ecosystems is thus incalculable—life on Earth as we know it would not exist without them. Nonetheless, economists and policy makers routinely attempt to put a monetary value on key ocean services to highlight the importance in investing to protect the oceans, and their results are striking. A recent study has estimated very conservatively the value of the annual 'gross marine product' of ocean services, which we can think of as equivalent to a country's gross domestic product, at US$2.5 trillion, and the total asset base of the oceans at US$24 trillion—if the oceans were a country they would be regarded as a powerhouse economy, among the largest in the world.

Unfortunately, human activities have been affecting the oceans for many years. It is now clear that overfishing, habitat destruction, pollution, the spread of exotic species, and the emission of climate-changing greenhouse gases are causing significant changes and damage to the oceans and the life forms living within them. This is destroying their natural beauty and biodiversity and severely eroding their ability to provide the life support systems and services that sustain our well-being and prosperity. Since the human population will increase from 7.7 billion to about 9.8 billion over the next thirty years, such pressures will intensify and pose an increasingly serious threat to human welfare.

Fortunately, there is rapidly growing community awareness, concern, and engagement about marine environmental matters, with the result that many more people worldwide are showing heightened interest in the science of marine biology and are seeking a greater understanding of our impacts on the oceans. This book aims to contribute to this auspicious trend by providing an accessible, up-to-date, and comprehensive introduction to the oceanic environment and the nature of life in the oceans so that readers can better appreciate the inherent beauty and complexity of marine systems, their significance to our planet and to human society, some of the consequences of increasing human impacts, and, ultimately, some of the actions required to put us on a path to a more sustainable relationship with our oceans so that they can be restored and protected for future generations.

# Chapter 1
# The oceanic environment

Viewed from space, our planet is clearly dominated by its greatest natural feature—a vast, deep, and interconnected mass of seawater—the Global Ocean. The Global Ocean has an area of about 362 million square kilometres, an average depth of 3,682 metres, and contains an enormous amount of water—about 1.34 billion cubic kilometres, which constitutes about 97 per cent of all the water that exists on our planet. As the science writer Arthur C. Clarke observed: 'How inappropriate to call this planet Earth, when clearly it is Ocean.'

## Geography of the Global Ocean

The Global Ocean has come to be divided into five regional oceans—the Pacific, Atlantic, Indian, Arctic, and Southern oceans, the latter extending from the coast of Antarctica to the line of latitude at 60° S (see Figure 1). Many of the marginal areas of these regional oceans are familiarly known as seas, for example the Caribbean Sea or the Red Sea.

Oceans are large, seawater-filled basins that share characteristic structural features (see Figure 2). The edge of each basin consists of a shallow, gently sloping extension of the adjacent continental land mass and is termed the continental shelf. Continental shelves typically extend offshore to depths of a couple of hundred

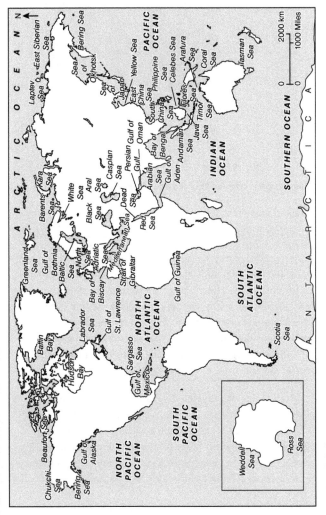

1. The Global Ocean.

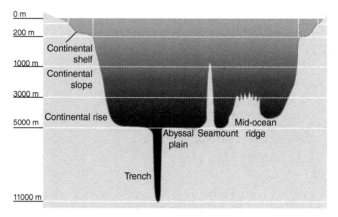

**2. Diagrammatic cross-section of an ocean basin.**

metres and vary from several kilometres to hundreds of kilometres in width.

At the outer edge of the continental shelf, the ocean floor drops off abruptly and steeply to form the continental slope, which extends down to depths of 2–3 kilometres. The continental slope then gives way to a more gently sloping continental rise that descends another kilometre or so to merge with a vast expanse of flat, soft, ocean bottom—the abyssal plain—that extends over depths of about 4–6 kilometres and accounts for about 76 per cent of the Global Ocean floor.

The abyssal plains are transected by extensive mid-ocean ridges—underwater mountain chains created by intense volcanic activity—that rise thousands of metres above the surrounding abyssal plains. Mid-ocean ridges form a continuous chain of mountains that extend linearly for 65,000 kilometres across the floor of the Global Ocean basins—akin to the seams on a baseball.

In some places along the edges of the abyssal plains the ocean bottom is cut by narrow trenches that plunge to extraordinary depths—3–4 kilometres below the surrounding ocean floor—and are thousands of kilometres long but only tens of kilometres wide. To provide a comparative sense of scale, the Grand Canyon averages about 1.6 kilometres deep, is 446 kilometres long, and is about 16 kilometres wide on average. The deepest known part of the Global Ocean—at around 11 kilometres below sea level—is at the bottom of one such trench, the Mariana Trench, located off Japan and the Philippine Islands.

Seamounts are another distinctive feature of ocean basins. They are typically extinct submarine volcanoes that rise 1,000 or more metres above the surrounding ocean floor but do not reach the surface of the ocean. Their peaks are thus hundreds to thousands of metres below the ocean surface. Seamounts generally occur in chains or clusters in association with mid-ocean ridges, although some rise from the ocean floor as solitary features. The Global Ocean contains an estimated 100,000 or so seamounts that rise more than 1 kilometre above the surrounding deep-ocean floor, and around 13,000 seamounts that rise more than 1.5 kilometres.

## Ocean habitats

Marine organisms live throughout the Global Ocean, from its sunlit surface to the bottom of its deepest trenches. They can be living in the open ocean, termed the pelagic zone, or in association with the ocean bottom, the benthic zone. Tiny organisms living suspended in the pelagic zone are known as plankton. Phytoplankton are those planktonic organisms capable of making their own food by photosynthesis; while zooplankton are small planktonic animals. Larger animals that actively swim through the water are referred to as nekton.

This colonization of every part of the marine environment by living things is now taken for granted, although in the 19th century it was widely believed that no life existed in what was then termed the 'azoic zone'—any part of the oceans beneath 300 fathoms (about 550 metres). Here the environment was considered a dead zone of darkness, too inhospitable for any life form. This notion was firmly laid to rest following the historic expedition of HMS *Challenger* (1872–6), the first ship to comprehensively explore the deeper parts of ocean basins and to discover marine life to depths of close to 6,000 metres.

We now know that the oceans are literally teeming with a huge diversity of life. Viruses, the most primitive of life forms, are astoundingly abundant, occurring at concentrations of tens of billions per litre of seawater; bacteria occur at concentrations of a billion or more per litre; phytoplankton at tens of millions per litre; zooplankton in the thousands per litre; and many hundreds of thousands of species of invertebrates, fish, mammals, and reptiles live in large numbers throughout the Global Ocean.

So what sort of environment does this plethora of life occupy?

## Salinity

The water in the oceans is in the form of seawater, a dilute brew of dissolved ions, or salts. Chloride and sodium ions are the predominant salts in seawater, along with smaller amounts of other ions such as sulphate, magnesium, calcium, and potassium (see Table 1).

**Table 1. Major ions in seawater**

| Ion | Weight (grams/kilogram of seawater) |
| --- | --- |
| Chloride ($Cl^-$) | 19.35 |
| Sodium ($Na^+$) | 10.76 |
| Sulphate ($SO_4^{2-}$) | 2.71 |
| Magnesium ($Mg^{2+}$) | 1.29 |
| Calcium ($Ca^{2+}$) | 0.41 |
| Potassium ($K^+$) | 0.40 |
| **Total** | **34.92** |

The total amount of dissolved salts in seawater is termed its salinity. Seawater typically has a salinity of roughly 35—equivalent to about 35 grams of salts in one kilogram of seawater. But this can vary, particularly in partially enclosed bays subject to high rates of evaporation, which increases salinity, or to freshwater inflow in the form of rain, river run off, or ice melt, which depresses salinity.

## Temperature

Most marine organisms are exposed to seawater that stays within a reasonably moderate temperature range compared to the extremes characteristic of terrestrial environments. Surface waters in tropical parts of ocean basins are consistently warm throughout the year, ranging from about 20–7°C, and up to around 30°C in shallow tropical bays at the height of summer. On the other hand, surface seawater in polar parts of ocean basins can get as cold as –1.9°C.

Ocean temperatures typically decrease with depth, but not in a uniform fashion. A distinct zone of rapid temperature transition is often present that separates warm seawater at the surface from cooler deeper seawater. This zone is called the thermocline layer (see Figure 3).

In tropical ocean waters the thermocline layer is a strong, well-defined, and permanent feature. It may start at around 100 metres and be a hundred or so metres thick. Ocean temperatures above the thermocline can be a tropical 25°C or more, but only 6–7°C just below the thermocline. From there the temperature drops very gradually with increasing depth. Thermoclines in temperate ocean regions are a more seasonal phenomenon, becoming well established in the summer as the sun heats up the surface waters, and then breaking down in the autumn and winter. Thermoclines are generally absent in the polar regions of the Global Ocean.

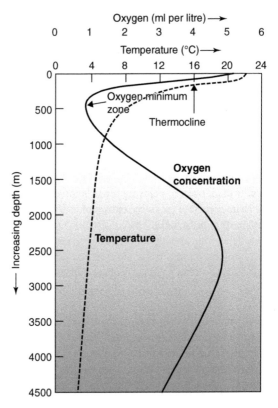

**3. Typical profile of the tropical ocean showing a thermocline layer and the oxygen minimum zone.**

Human-induced global climate change, which is resulting in increasing average global air temperatures, is also resulting in rising ocean temperatures. The oceans absorb nearly all the additional warmth created from human greenhouse gas emissions and the resulting greenhouse effect; over the last forty years the oceans have absorbed an astonishing 93 per cent of the additional heat created. This oceanic heat sink has thus played a major role in shielding humanity from climate change by moderating the rise in atmospheric temperatures that would have occurred otherwise. To put this into perspective, if the heat generated between 1955 and 2010 had all gone into the Earth's atmosphere instead of its oceans, average global surface air temperatures would have increased by about 36°C, instead of the 1.1°C increase thus far.

As a result, however, the surface waters of most of the Global Ocean are now close to 1°C warmer than 140 years ago and in some places more than 3°C warmer. Later in this century, temperatures will begin to increase in deeper parts of the oceans as well as the warmer surface seawater is slowly mixed to deeper depths. This warming trend is causing a rapid decrease in the thickness and coverage of sea ice in the Arctic Ocean and the thinning of ice shelves in Antarctica. It is also impacting on marine organisms and the functioning of marine ecosystems in a variety of important ways that we will explore throughout this volume.

## Light

The amount of sunlight that strikes the surface of the oceans varies considerably with time of day, cloud cover, time of year, and latitude. The depth to which this available sunlight manages to penetrate the surface of the oceans (the sunlit layer or photic zone of the oceans) depends largely on the amount of suspended particles in the seawater. These consist of a mixture of suspended sediment and living and dead organic matter. As a rule of thumb, light does not penetrate much beyond about 150–200 metres in most parts of the Global Ocean, with red light being absorbed within the first few metres and green and blue light penetrating the deepest. In temperate coastal seas light may penetrate only a few tens of metres at certain times of year because of the large amounts of particulates in the seawater.

## Pressure

Pressure is a defining feature of the marine environment. In the oceans, pressure increases by an additional atmosphere every 10 metres (one atmosphere of pressure is roughly equivalent to the air pressure at sea level). Thus, an organism living at a depth of 100 metres on the continental shelf experiences a pressure ten times greater than an organism living at sea level; a creature living at 5 kilometres depth on an abyssal plain experiences pressures some 500 times greater than at the surface; those organisms dwelling in the deeper parts of oceanic trenches are subject to pressures some 1,000 times greater than a sea-level dweller—the pressure at these depths equates to a massive 10,000 tonnes per square metre.

## Oxygen

Dissolved oxygen is reasonably abundant throughout most parts of the Global Ocean. However, the amount of oxygen in seawater is much less than in air—seawater at 20°C contains about 5.4 millilitres of oxygen per litre of seawater, whereas air at this temperature contains about 210 millilitres of oxygen per litre. The colder the seawater, the more oxygen it contains; for example, seawater at 0°C contains around 7.8 millilitres of oxygen per litre.

Oxygen is not distributed evenly with depth in the oceans. Oxygen levels are typically high in a thin surface layer 10–20 metres deep. Here oxygen from the atmosphere can freely diffuse into the seawater, plus there are abundant phytoplanktonic organisms in the photic zone producing oxygen through photosynthesis. Oxygen concentration then often decreases rapidly with depth to reach very low levels, sometimes close to zero, at depths of around 200–1,000 metres. This region is referred to as the oxygen minimum zone (see Figure 3). This zone is created by the low rates of replenishment of oxygen diffusing down from the surface layer of the ocean, combined with the high rates of depletion of oxygen by bacterial decay of particulate organic matter sinking from the surface.

Beneath the oxygen minimum zone, oxygen content increases with depth such that the deep oceans contain quite high levels of oxygen, though not generally as high as in the surface layer. The higher levels of oxygen in the deep oceans reflect in part the origin of deep-ocean seawater masses, which are derived from cold, oxygen-rich seawater that sinks rapidly down from the surface of polar oceans, thereby conserving its oxygen content. Also, compared to life in near-surface waters, organisms in the deep ocean are comparatively scarce and have low metabolic rates, thus consuming little of the available oxygen.

## Carbon dioxide and ocean acidification

In contrast to oxygen, carbon dioxide ($CO_2$) dissolves readily in seawater. Most of it is then converted into carbonic acid ($H_2CO_3$), bicarbonate ion ($HCO_3^-$), and carbonate ion ($CO_3^{2-}$), with the proportion of all four forms existing in a complex equilibrium, as shown in the following equation:

$$CO_2 + H_2O \leftrightarrow H_2CO_3 \leftrightarrow H^+ + HCO_3^- \leftrightarrow H^+ + CO_3^{2-}$$

Bicarbonate ion ($HCO_3^-$) is by far the dominant form in seawater, with the other forms occurring in much smaller amounts. However, changes in $CO_2$ concentration affect this equilibrium and, hence, the pH of seawater. For example, if more $CO_2$ is added to seawater some of the available carbonic acid ($H_2CO_3$) loses an $H^+$, which lowers the pH and creates more bicarbonate ion ($HCO_3^-$). Furthermore, some of this additional $H^+$ reacts with some of the available carbonate ion ($CO_3^{2-}$), reducing its availability. In contrast, removing $CO_2$ creates more carbonic acid ($H_2CO_3$), thus binding more $H^+$, increasing the pH, and making more carbonate ion ($CO_3^{2-}$) available.

Seawater is naturally slightly alkaline, with a pH ranging from about 7.7 to 8.2, and marine organisms have become well adapted to life within this range. Seawater near the surface of the oceans is generally at the higher end of the pH range because abundant photosynthetic organisms in the photic zone are taking up carbon dioxide. Furthermore, surface seawater is generally warmer than deep-ocean water and the warmer the seawater, the less carbon dioxide it can absorb. In deep, colder parts of the oceans, where no photosynthesis is taking place, carbon dioxide concentrations are higher, and the pH is often at the lower end of the range.

As a result of the carbon dioxide–carbonic acid–bicarbonate–carbonate equilibrium in seawater, the Global Ocean is a vast reservoir of inorganic carbon, which has important implications for marine life and human society. From a biological perspective, carbon is never a limiting factor for the growth of marine photosynthetic organisms, as it is for terrestrial plants. From a planetary perspective, the Global Ocean is an enormous natural sink for atmospheric carbon dioxide, a climate-changing greenhouse gas.

At present, the Global Ocean is absorbing at least 25 per cent of the roughly 41 billion tonnes of carbon dioxide being spewed into the atmosphere each year by humans burning fossil fuels and deforesting the planet's surface—this is roughly 1.2 million tonnes of carbon dioxide per hour. Another 25 per cent or so is absorbed by forests, with the balance accumulating in the atmosphere. The net result is that atmospheric carbon dioxide concentrations are currently rising at a rate of about 3 parts per million (ppm) per year, which is why carbon dioxide concentrations in the planet's atmosphere have increased from a pre-industrial level of 278 ppm to a level now greater than 405 ppm and rising rapidly. It is a sobering thought that if it were not for the Global Ocean's uptake of large amounts of anthropogenic carbon dioxide, atmospheric carbon dioxide concentrations would now be over 460 ppm.

The Global Ocean thus plays a fundamental role in damping the rate of human-induced climate change not only by absorbing excess heat from the atmosphere, but also by mopping up much of the excess carbon dioxide spewed into the atmosphere since the Industrial Revolution. Unfortunately, all this absorbed carbon dioxide is starting to change the basic chemistry of the Global Ocean, making it more acidic on average, a process called ocean acidification.

The average pH of surface seawater in 1870 was about 8.18. It is now around 8.0 to 8.1 and by 2100 is predicted to be about 7.7 to 7.8 if humans continue to emit anthropogenic carbon dioxide at current rates. These seem like small changes numerically but, because the pH scale is logarithmic, they equate to about a 30 per cent and a 170 per cent increase in acidity, respectively. This rate of change is about ten times faster than anything that has occurred over the last 65 million years. As acidified surface seawater gradually mixes with deeper water, the entire Global Ocean is affected.

Ocean acidification is already causing the pH of many regions of the Global Ocean to decline to levels below the range naturally experienced by marine organisms. Surface water pH values less than 8.0 are now commonly measured and values as low as 7.6 have been measured in some regions. Declining ocean pH is now affecting many types of marine organisms, particularly those with external shells or internal skeletons containing calcium carbonate, such as corals, clams, oysters, sea urchins, starfish, and some species of plankton and algae. These organisms use carbonate in seawater to manufacture their shells or skeletons. As noted above, as pH decreases, there is less carbonate in seawater, which makes it more difficult for these organisms to build proper shells or skeletons and grow. Furthermore, as carbonate depletion increases, seawater begins to draw carbonate out of the shells and skeletons of these organisms, corroding and weakening them, with obvious impacts on their health.

Ongoing research is revealing many examples of the negative effects of ocean acidification on marine life. Effects have already been observed in foraminifera—microscopic organisms which are abundant in the plankton and which build calcium carbonate shells. The shells of foraminifera from the Southern Ocean are now significantly thinner compared to preserved specimens from pre-industrial times. Studies on reef corals have shown that ocean acidification significantly reduces the ability of some species to produce their skeletons, which impacts their growth and ability to recover from other environmental stresses. It has also been shown that the shells of sea butterflies, or pteropods—tiny planktonic sea snails—dissolve at ocean pH levels predicted to occur in 2100. Sea butterflies are particularly important in polar and subpolar oceans where they are eaten by whales, seabirds, and commercial fishes such as salmon, herring, cod, and mackerel. A comprehensive review of the sensitivity of marine animals to ocean acidification has found that a significant proportion of the 153 different species of marine invertebrates and fishes studied exhibited negative effects of some kind. This suggests that ocean acidification will cause substantial changes to ocean ecosystems within this century, including permanent shifts in species composition.

## The oceans in motion

On a planetary scale, the surface of the Global Ocean is moving in a series of five enormous, roughly circular, wind-driven current systems, or gyres, each thousands of kilometres in diameter (see Figure 4). The northern hemisphere gyres in the North Pacific and North Atlantic Oceans flow clockwise; the southern hemisphere gyres in the South Pacific, South Atlantic, and the Indian Oceans flow counterclockwise. These gyres transport enormous volumes of water and heat energy from one part of an ocean basin to another and carry along many kinds of plankton.

The North Atlantic Ocean gyre provides a good example of the dynamics of gyre systems. Here the surface waters circulate

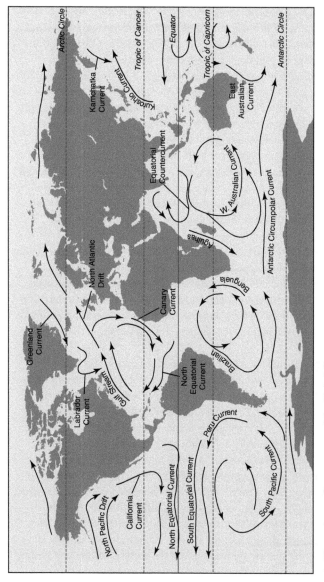

4. **The major surface currents of the Global Ocean.**

around a stable centre known as the Sargasso Sea. The northward-flowing western edge of this gyre comprises the Gulf Stream. The Gulf Stream is a 50–75-kilometre-wide fast-moving surface current that transports vast volumes of warm, salty tropical seawater at speeds averaging 3–4 kilometres per hour up along the eastern edge of the North American continent. This warm current leaves the coast of North America at around South Carolina and traverses the North Atlantic as the North Atlantic Drift Current, releasing its heat into the atmosphere along the way. The gyre then turns southward and meanders down along the western edge of Europe and Africa as the cooler, broader, and more slowly moving Canary Current. This current then curves westward along the equator to form the North Equatorial Current which flows into the Caribbean region to complete the gyre.

Beneath the surface, the deeper water masses of the Global Ocean are also in motion. This motion is not created by wind, as for surface currents, but by buoyancy changes in seawater occurring in the polar oceans. This results in a more stately flow referred to as thermohaline circulation because the buoyancy changes are a result of changes in the temperature and salinity of the seawater.

The Atlantic Ocean provides a good example of thermohaline circulation. The Gulf Stream transports large amounts of warm, salty seawater from the tropics to polar latitudes in the North Atlantic Ocean. Here it is cooled and its salinity increased by the addition of salt extruded from the freezing of Arctic seawater. This process creates cold, salty, and dense seawater that sinks rapidly to great depths. Large amounts of such seawater are formed and sink in the seas off Greenland, Iceland, and Norway forming what is called the North Atlantic Deep Water (see Figure 5). From there this water mass flows slowly southward near the bottom of the Atlantic Ocean basin to upwell hundreds of years later near the coast of Antarctica. Similarly, very cold, salty, and very dense seawater is created off the coast of Antarctica and sinks

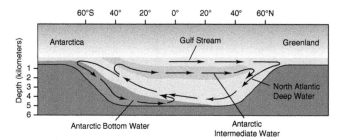

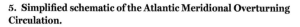

**5. Simplified schematic of the Atlantic Meridional Overturning Circulation.**

to the bottom of the Atlantic basin and flows northwards beneath the North Atlantic Deep Water and penetrates far into the North Atlantic basin. This mass of water is referred to as the Antarctic Bottom Water. Some of the upwelling North Atlantic Deep Water diverges towards Antarctica, cools, and merges with Antarctic Bottom Water to sink back into the depths. However, another portion of this upwelling seawater diverges away from Antarctica and moves back towards the equator where it becomes somewhat warmer and less saline and sinks beneath the Gulf Stream water to create another water mass of intermediate density called the Antarctic Intermediate Water that penetrates well into the North Atlantic. This Atlantic Ocean circulatory system is termed the Atlantic Meridional Overturning Circulation (AMOC).

The AMOC plays an important role in transporting heat northwards from equatorial regions and releasing it into the atmosphere above the North Atlantic Ocean. It thus has a major influence on northern hemisphere climate patterns. Oceanographers have speculated that human-induced climate change could affect this system by adding more low-salinity water to the polar North Atlantic because of enhanced melting of sea ice and the Greenland Ice Sheet. This would make the surface waters more buoyant, decreasing the rate of sinking of seawater, and weakening the AMOC.

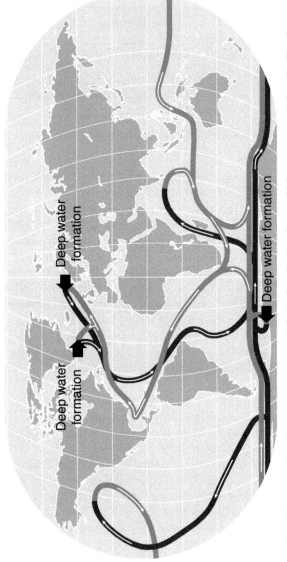

6. Simplified view of the Great Ocean Conveyor Belt. Darker shading shows cold deep currents; Lighter shading shows warmer shallow currents.

There is mounting evidence that this scenario may be playing out. Recent studies have shown that the AMOC flow rate is now about 15 per cent slower than at any time in the past 1,600 years. It is not yet completely clear to what extent this weakening is being driven by natural climate variability or human-induced climate change. It is likely that both mechanisms are involved, with anthropogenic warming working together with natural climate variability to sustain or enhance ice melting and AMOC weakening. Further work is needed to better predict the state of the AMOC over the next half-century since further weakening or, in the worst case, collapse of the system, would lead to profound changes in climate patterns throughout the northern hemisphere.

The AMOC is part of a much larger, planetary-scale ocean circulation system. The sinking of seawater off Greenland, Iceland, and Norway, supplemented by the sinking of seawater off Antarctica, drives a remarkable Global Ocean current system known as the Great Ocean Conveyor Belt. Like a giant conga line, the Great Ocean Conveyor flows through and links up the Atlantic, Indian, Pacific, and Southern Oceans (see Figure 6). Simply put, the Conveyor moves cold saline water in the form of a deep current from the Atlantic Ocean into the Southern Ocean where it then flows in a clockwise direction around Antarctica, giving off branches into the Indian and Pacific Oceans. These currents rise to the surface, warm, and return as surface currents back across the Pacific, Indian, and Atlantic Oceans to the starting point of the Conveyor system in the polar North Atlantic, a process that takes about 1,000 years. The Great Ocean Conveyor moves at much slower speeds than the wind-driven surface currents—a few centimetres per second—but it moves enormous volumes of water—more than a hundred times the flow of the Amazon River. It is thus akin to a planetary-scale distribution system, transporting oxygen, nutrients, and heat throughout the oceans of the planet and moderating the global climate.

# Chapter 2
# Marine biological processes

Marine microbes—microscopic, single-celled organisms—are much more diverse and abundant in the oceans than previously thought and play key roles in the production and flow of organic matter and energy and the cycling of nutrients in the oceans. Roughly half of the planet's primary production—the synthesis of organic matter by chlorophyll-bearing organisms using light energy from the sun—is produced within the Global Ocean. On land the primary producers are large, obvious, and comparatively long-lived life forms—the trees, shrubs, grasses, and food crops characteristic of the terrestrial landscape. The situation is quite different in the oceans where, for the most part, the primary producers are phytoplanktonic microbes suspended in the sunlit surface layer of the oceans. These energy-fixing microorganisms—the marine environment's invisible pasture—form the basis of the marine food web, the network of pathways through which food energy is transferred to all the other organisms in the marine system including other microbes, zooplankton, fish, marine mammals, and, ultimately, humans.

## An ocean of microbes

Microbes are astoundingly abundant in the pelagic marine environment. If one added up the weight of all the microbes in the oceans it would account for more than two-thirds of the total marine living biomass. The oceans are undoubtedly a vast, clear soup of microbial life.

The revolution in our understanding of the importance of microbes in marine processes began in the 1970s when improved microscopy and counting techniques revealed the unsuspected diversity and extraordinary abundance of the marine microbial world, or marine microbiome. Since then knowledge of the structure and function of the marine microbiome has advanced much further due to more intensive sampling of the oceans and a convergence of new technologies that include: further improvements in microscopic imaging; the recent development of metagenomic approaches that allow the genomes of huge numbers of microbes sampled from the natural environment to be quickly analysed by very rapid sequencing of their genetic material; and a revolution in bioinformatics based on advances in computing that assist in the analysis and interpretation of the enormous amounts of metagenomic data being generated.

Several 21st-century ocean expeditions, reminiscent of the classic oceanographic voyages of discovery of the 19th century, have played key roles in marine microbiome research. The *Tara* Oceans Expedition is one such voyage. The *Tara* is a 36-metre schooner that sailed the world's oceans between 2009 and 2013 collecting more than 35,000 seawater samples from the surface to 1,000 metres for metagenomic analysis of the microbes they contained. The work so far has revealed a genetic goldmine in the oceans—an inventory of around 40 million genes, over 80 per cent of which are new to science, reflecting the incredible diversity of the marine microbiome.

## Marine microbial diversity

Four main groups of microbes live in the oceans—bacteria, archaea, protists, and viruses (see Figure 7).

### Bacteria

Prokaryotic – cell without a nucleus

Less than 2 µm

1 billion to 10 billion in a litre of seawater

Autotrophic, heterotrophic or mixotrophic nutrition

### Archaea

Prokaryotic – cell without a nucleus

Similar in size and shape to bacteria but with a different biochemistry

Abundant component of marine microbiome

Utilise many different sources of energy

### Protists

Eukaryotic – cell contains a nucleus

1 – 200 µm in size

1 million to 100 million in a litre of seawater

Morphologically and genetically very diverse

Autotrophic, heterotrophic, mixotrophic and parasitic forms

### Viruses

0.02 – 0.3 µm in size

10 billion to 100 billion in a litre of seawater

Infect bacteria, archaea and protists

**7. Marine microbial diversity.**

Marine Biology

Bacteria form part of a group of life forms known as prokaryotes—single-celled organisms without a nucleus. Marine bacteria are minute, generally less than 2 μm in diameter. They can be non-motile or have one or more whip-like appendages, called flagella, which provide them with a degree of mobility within the microworld within which they feed, grow, and reproduce. Densities of bacteria in seawater range from about a billion up to a staggering 10 billion per litre. Due to their microscopic size, a billion marine bacteria weigh only about 0.1 mg and would occupy a scant 0.0000001 per cent of a litre of seawater. Thus, despite being incredibly numerous, bacteria occupy very little space in the oceans.

Some marine bacteria are autotrophic—their cells contain chlorophyll allowing them to produce their own food by photosynthesis. Others are heterotrophic—they obtain nutrition by absorbing organic molecules dissolved in seawater (Dissolved Organic Matter or DOM) through their cell membranes, or by colonizing and absorbing energy from tiny particles of organic matter suspended in seawater (Particulate Organic Matter or POM). Some are mixotrophic—able to combine autotrophic and heterotrophic modes of nutrition. Marine bacterial cells have extremely high metabolic rates and can grow and reproduce asexually by cell division very quickly, often daily, sometimes in a matter of hours. Bacterial populations thus have the potential to produce huge population 'blooms' very rapidly under favourable conditions.

Archaea, like bacteria, are prokaryotes and look superficially like bacteria in size and shape, but their basic biochemistry is very different. Archaea were first discovered in harsh environments, such as hot springs and very saline lakes, and were originally considered to be 'extremophiles', only found in such extreme environments. They are now known to be abundant in many other habitats, including the pelagic zone of the oceans, where they comprise an abundant and diverse component of the microbial biomass.

Protists are eukaryotic, meaning their cells possess a nucleus, but they are single-celled, not multicellular like other eukaryotic organisms. Marine protists range in size from about 1 to 200 μm and occur at densities of 1 million to 100 million in a litre of seawater. The protists are the most diverse group of organisms in the oceans, much more diverse than marine bacteria and marine animals. Metagenomic analysis suggests that there are hundreds of thousands of different types of protists in the oceans, most known only by their genetic signature. Despite being single-celled, marine protists have evolved a dazzling array of shapes and specialized sub-cellular structures. Some have cilia or flagella for locomotion, while others are amoeboid; some have internal skeletons, while others produce external shells; and many have developed specialized structures for capturing food and sensing the environment. Marine protists can be autotrophic or heterotrophic while some are mixotrophic. The heterotrophs are predators of bacteria and other protists and are often referred to as protozoa. Some protists live as parasites or symbionts in association with other organisms.

Viruses are by far the most abundant 'life forms' in the oceans—existing at the boundary between living and non-living entities. They are much smaller than bacteria, with a diameter typically between 0.02 and 0.3 μm and are present at astonishing densities of 10 to 100 billion in a litre of seawater.

A virus consists of a protein coat encapsulating a small amount of genetic information in the form of nucleic acids. Viruses are not self-replicating and must infect a host organism to survive and propagate. In the oceans, bacteria, archaea and protists are readily available hosts. Viruses that infect such hosts are called phages and phage infections ultimately cause the death and break-up of the host cells, releasing vast numbers of identical copies of the original virus, along with the cell's remains, into the seawater. Phage infections are thus major contributors to a large store of DOM and POM in the oceans.

# Marine phytoplankton diversity

Autotrophic bacteria are a key component of marine phytoplankton, contributing in the order of 30–50 per cent of total marine primary production. One group, the cyanobacteria, or blue-green bacteria, are particularly important. One type of cyanobacterium, *Prochlorococcus*, is one of the smallest (a sphere about 0.6 μm in diameter) and most numerous photosynthesizing microbes in the oceans (see Figure 8(a)). Because of its small size, this microbe was only discovered in the 1980s. We now know that it is incredibly abundant in tropical and sub-tropical parts of the Global Ocean between 40° N and 40° S latitudes from the surface to about 200 metres at densities of over 100 million per litre of seawater. It has been calculated that the oceans contain about $3 \times 10^{27}$ (or three billion billion billion) *Prochlorococcus* cells, probably making this bacterium the most abundant photosynthetic organism on the planet and responsible for about 5 per cent of global primary production. Gene sequencing has revealed that *Prochlorococcus* comprises thousands of ecotypes—each ecotype being a genetically distinct strain adapted to a specific environment. For example, there is an ecotype adapted to high light levels in surface waters and another adapted to low light levels in deeper waters.

Protists comprise the other major group of marine phytoplankton. Diatoms are an important component of this group and are very abundant, particularly in polar waters and upwelling regions. Their cells are generally about 10 to 200 μm in size, making them much larger than photosynthetic bacteria. Each diatom cell is enclosed within an ornately sculptured, clear glass box made of silica, called a frustule (see Figure 8(b)). The individual cells of some species of diatom can link up to form colonies of long chains.

The dinoflagellates and silicoflagellates are two other significant types of protistan phytoplankton. Their cells range in size from

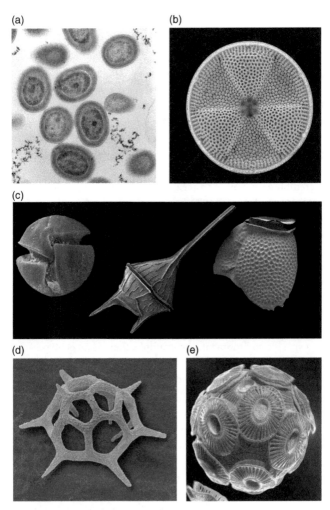

8. Marine phytoplankton diversity: (a) Transmission electron micrograph of cells of the cyanobacterium *Prochlorococcus*; (b) Scanning electron micrograph of a diatom; (c) Scanning electron micrograph of several types of dinoflagellates; (d) Scanning electron micrograph of the skeleton of a silicoflagellate; (e) Scanning electron micrograph of a coccolithophore.

about 2 to 20 μm. Dinoflagellate cells possess two hair-like flagella, providing them with some limited motility, and are often armoured with translucent plates made of cellulose (see Figure 8(c)). Silicoflagellates possess one long flagellum and an internal skeleton made of silica (see Figure 8(d)).

Coccolithophores are yet another important kind of protistan phytoplankton. Their cells are in the 20–200 μm size range. Each cell is covered with small ornamented plates, called coccoliths, made of calcium carbonate (see Figure 8(e)).

## Factors affecting marine primary production

Phytoplankton use the chlorophyll pigments in their cells to harvest the energy of sunlight penetrating the photic zone of the oceans. Through the process of photosynthesis this energy is used to synthesize energy-rich carbon-containing organic compounds, such as glucose. Dissolved carbon dioxide ($CO_2$) in the seawater provides the source of inorganic carbon for this process. Oxygen ($O_2$) is a by-product of photosynthesis and is released into the surrounding seawater.

Photosynthesis is a complex, multistep process, but it can be summarized with the following simplified equation:

$$6CO_2 + 6H_2O \xrightarrow{\text{Sunlight}} \underset{\substack{\text{organic} \\ \text{compounds}}}{C_6H_{12}O_6} + 6O_2$$

The organic matter produced by phytoplankton is the energy base, or first trophic level, of the Global Ocean; it represents the major source of energy supporting life in the oceans. The energy at the first trophic level is used by a diversity of marine organisms at the second trophic level—the primary consumers or herbivores—which are in turn eaten by consumers at higher trophic levels in the system.

The rate of photosynthesis, and hence primary production, decreases with depth in the oceans because of decreasing light intensity. Since the upper layers of the oceans are a naturally turbulent environment, phytoplankton are mixed to various depths within the water column depending on the strength of vertical circulation. If phytoplankton are to grow and reproduce, they must spend enough time above a certain depth in the photic zone, often referred to as the 'critical depth', to be able to photosynthesize more energy than is required for their basic metabolic requirements. Otherwise all the energy produced is respired and there is nothing left for growth. Thus, light availability and the strength of vertical mixing are important factors limiting primary production in the oceans.

Nutrient availability is the other main factor limiting the growth of primary producers. One important nutrient is nitrogen, which phytoplankton require for a variety of metabolic functions. For example, nitrogen is a key component of amino acids—the building blocks of proteins. Nitrogen is absorbed by most marine photosynthetic organisms in the form of dissolved ammonium ($NH_4^+$), nitrite ($NO_2^-$), or nitrate ($NO_3^-$). Nitrogen 'fixing' bacteria, such as the cyanobacterium *Trichodesmium*, provide a source of these essential inorganic nitrogen compounds in the oceans. These bacteria have the specialized ability to convert or 'fix' molecular nitrogen ($N_2$) dissolved in the seawater into ammonia ($NH_3$), which is then used by their cells to synthesize nitrogen-rich organic compounds such as proteins. When nitrogen-fixing bacteria die, these organic compounds are released into the seawater where other kinds of bacteria break them down and recycle them back into inorganic forms of nitrogen which are then available to other primary producers.

Photosynthetic marine organisms also need phosphorus, which is a requirement for many important biological functions, including the synthesis of nucleic acids, a key component of DNA. Phosphorus in the oceans comes naturally from the erosion of rocks and soils on land and is transported into the oceans by rivers, much of it in the form of dissolved phosphate ($PO_4^{3-}$), which can be readily absorbed by marine photosynthetic organisms.

Inorganic nitrogen and phosphorus compounds are most abundant in the deep ocean where bacteria decompose and recycle a rain of dead organic material sinking down from the surface waters back into inorganic forms of nitrogen and phosphorus. When the upper layer of the ocean is well mixed, or unstratified, these deep nutrient-rich waters are mixed up into the photic zone, bringing an abundant supply of nutrients to the surface. But when a thermocline is present, it acts as a barrier to the regeneration of nutrients from the deep, oceanic waters below. Under such circumstances, and if light levels are not limiting, photosynthetic organisms will rapidly deplete nutrients from the surface layer above the thermocline. In practice, inorganic nitrogen and phosphorus compounds are not used up at the same rate. Thus, one will be depleted before the other and becomes the limiting nutrient at the time, preventing further photosynthesis and growth of marine primary producers until it is replenished. Nitrogen is generally considered to be the rate-limiting nutrient in most oceanic environments, particularly in the open ocean.

Marine primary producers also require the micronutrient iron, which is used in a variety of metabolic processes essential for photosynthesis and growth. The iron in the oceans is derived from iron-rich dust that is blown far out into the oceans from deserts during dust storms. Iron deposits at the edge of continents are another source. Adequate concentrations of dissolved iron exist in most parts of the Global Ocean, so it is not normally a limiting factor for primary production. Nevertheless, in some open-ocean regions, such as the equatorial and North Pacific Ocean and large parts of the Southern Ocean, concentrations of dissolved iron are so low that iron becomes the rate-limiting factor for primary production despite high levels of nitrogen and phosphorus being present. Such areas are referred to as high nutrient–low chlorophyll (HNLC) regions, the low chlorophyll concentrations reflecting the dearth of chlorophyll-containing organisms in the seawater.

Some researchers have suggested that artificially fertilizing HNLC regions with iron to kick-start primary production in these otherwise nutrient-rich regions could be a way of mitigating climate change. The idea is that the resultant phytoplankton bloom will draw down large amounts of carbon dioxide from the atmosphere. When the phytoplankton die, they will sink into the deep ocean where, if they reach the ocean floor, the carbon in their tissues will be potentially locked away into marine sediments. In other words, if done on a large enough scale, these artificially created blooms of phytoplankton might enhance the naturally occurring phytoplankton mediated 'pumping' of carbon dioxide from the atmosphere into the ocean interior and potentially into long-term storage in deep-ocean sediments. Indeed, some corporations have plans to convert theory into practice and have proposed to undertake large-scale iron fertilization of the oceans and earn income from the creation and sale of carbon credits in global carbon trading markets.

Over a dozen, small-scale (about 100 km$^2$) trials involving fertilizing patches of the open ocean with around a tonne of dissolved iron dispersed from research vessels have shown that primary production can be stimulated in this way over periods of a few days to weeks. Whether this will ever be a practical way to 'geoengineer' the large-scale removal of carbon dioxide from the atmosphere is very uncertain. To work well and lock away significant amounts of carbon in marine sediments, the process will have to stimulate the right kind of phytoplankton blooms—large diatom cells that are resistant to being grazed by organisms at the second trophic level, and that sink rapidly to the sea floor. Otherwise the primary producers will be quickly consumed by zooplankton and the carbon released as carbon dioxide back into the surface waters. The trials so far have shown that it is very difficult to verify what amount of carbon, if any, is being transported to the ocean floor and whether the process would actually make a difference if adopted on a large scale. More importantly, no one can predict what the adverse effects might be on the greater marine biological system of industrial-scale ocean fertilization with iron. It would almost certainly change the composition and dynamics of phytoplankton communities on a large scale and impact marine food webs in unintended ways. Indeed, in 2008 the parties to the United Nations Convention on Biological Diversity agreed to a moratorium on all artificial ocean fertilization projects except for small ones in coastal waters. Thus, although research on iron fertilization is furthering our understanding of primary production processes in the oceans, it is unlikely that it will lead to a viable and safe technology that can be used to help limit human-induced climate change with minimal risks.

## Measuring marine primary productivity

The rate of primary production—termed primary productivity—varies considerably over space and time in the Global Ocean. Primary productivity is often expressed as the number of grams of carbon (C) 'fixed', or incorporated into organic matter, per square metre of ocean surface per year (g C m$^{-2}$ yr$^{-1}$). Unless otherwise stated, estimates of primary productivity refer to net primary productivity. This figure is of interest to marine biologists because it is a measure of the proportion of fixed carbon that is used for growth of phytoplankton and which is therefore available to higher trophic levels. It excludes the fixed carbon that is respired away as carbon dioxide to maintain cellular functions and which is therefore lost to the next trophic level.

Measuring primary productivity in the oceans is a challenging business. It can be done by putting samples of seawater into bottles, exposing them to light, and measuring the amount of oxygen released by the photosynthetic organisms in the bottles.

This can be converted into estimates of the amount of carbon that has been fixed into organic material because the number of molecules of oxygen produced is closely equivalent to the number of molecules of carbon dioxide fixed into organic matter (see the equation for photosynthesis, p. 30). Another approach is to measure more directly the amount of carbon incorporated into the phytoplankton in the bottles using radioactive carbon-14 as a tracer. Using these methods, estimates of primary productivity in a patch of ocean at a particular time can be obtained.

The problem with these *in situ* techniques is that, even with the most intensive sampling regime, they can only provide estimates of primary productivity for a small number of locations at any one time. It is therefore difficult to build up a global and dynamic picture of primary productivity using these approaches. This changed when satellite observations of the colour of the surface of the oceans started to become available from 1978 with the launch of the Nimbus 7 satellite. This satellite carried an instrument called the Coastal Zone Color Scanner which measured the wavelengths of light reflected from the ocean surface—basically, the greener the ocean surface, the greater the chlorophyll concentration, and hence the more photosynthetic organisms present. More advanced satellite-borne sensors are now used routinely to estimate chlorophyll concentrations, and thus primary productivity, over very large areas of the Global Ocean. Repeated satellite-derived measurements of the entire Global Ocean over scales of days, weeks, and years, coupled with direct, *in situ* measurements, have provided a greatly improved synoptic understanding of spatial and temporal patterns of global primary productivity.

## Global patterns of marine primary productivity

The overall pattern of primary productivity in the Global Ocean depends greatly on latitude (see Figure 9). In polar oceans primary production is a boom-and-bust affair driven by light

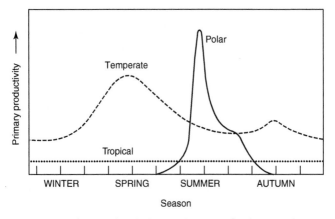

**9. Depiction of seasonal variation in primary productivity in polar, temperate, and tropical oceans.**

availability. Here the oceans are well mixed throughout the year, so nutrients are rarely limiting. But during the polar winter there is no light, and thus no primary production is taking place. In the spring, light levels and day length both increase rapidly, and a point is reached during the year in which both nutrients and light become simultaneously non-limiting, and an intense phytoplankton bloom commences. This may last for several months until light once again becomes limiting in the autumn. Although limited to a short seasonal pulse, the total amount of primary production can be quite high, especially in the polar Southern Ocean where annual productivity can be in the order of $100 \text{ g C m}^{-2} \text{ yr}^{-1}$ and in some areas much more.

In temperate open-ocean regions, primary productivity is linked closely to seasonal events. In the winter the ocean surface cools and the thermocline breaks down, assisted by strong winds that mix the surface layers. This allows surface waters to become well mixed with deeper, nutrient-rich seawater. However, light levels are low in the winter and limit primary productivity. In the spring, as the days lengthen, and the sun gets higher in the sky, a time is reached when both light and nutrients become non-limiting and a spring phytoplankton bloom is triggered. In the summer, although light is now abundant, a thermocline becomes re-established as the surface waters warm, locking out the photic zone from the nutrient-rich deeper waters. Nutrients are now limiting and the spring bloom 'crashes'. In the autumn, the thermocline breaks down once again and nutrients are regenerated into the photic zone. If this occurs early enough in the autumn, while there is still enough sunlight, then nutrients and light both become non-limiting for a short time, and an autumn phytoplankton bloom may occur. This bloom will persist until light once again becomes a limiting factor in the late autumn and winter. Although highly seasonal, primary productivity in temperate oceans totals in the order of 70–120 g C m$^{-2}$ yr$^{-1}$, levels similar to a temperate forest or grassland.

In tropical open oceans, primary productivity is low throughout the year. Here light is never limiting but the permanent tropical thermocline hinders the mixing of deep, nutrient-rich seawater with the surface waters. Hence nutrients are present at permanently low levels in the photic zone, which limits primary productivity. For this reason, open-ocean tropical waters are often referred to as 'marine deserts', with productivity generally less than about 30 g C m$^{-2}$ yr$^{-1}$, which is comparable to a terrestrial desert.

Some of the most productive marine environments occur in the coastal ocean above the continental shelves. This is the result of a phenomenon known as coastal upwelling which brings deep, cold, nutrient-rich seawater to the ocean surface, creating ideal conditions for primary productivity which can be more than 500 g C m$^{-2}$ yr$^{-1}$, comparable to a terrestrial rainforest or cultivated farmland. These hotspots of marine productivity are created by wind acting in concert with the planet's rotation.

The phenomenon can be explained in basic terms as follows. When a steady wind blows over the surface of the ocean it sets the surface layer moving in the direction of the wind. This top layer of moving seawater will in turn set a layer of seawater beneath it moving, although a bit more slowly than the layer above it, and so on down the water column until all the wind's energy has been transferred into moving water. However, due to the planet's rotation, which creates a phenomenon known as the Coriolis Effect, each of these moving layers of seawater curls slightly to the right in the northern hemisphere and to the left in the southern hemisphere. This creates a characteristic spiralling pattern of water movement down the water column called the Ekman Spiral. The net outcome of an Ekman Spiral is that the average direction of flow of the seawater set in motion by the wind is at roughly right angles to the direction of the surface wind—to the right of the wind direction in the northern hemisphere, and to the left in the southern hemisphere. This net movement of water to the right or left of the wind direction is known as Ekman transport.

Coastal upwelling can occur when prevailing winds move in a direction roughly parallel to a coastline to create offshore Ekman transport. Coastal upwelling is particularly prevalent along the west coasts of continents. In the southern hemisphere, when a steady wind blows roughly from the south along the western edge of a continent, the net movement of the top 100 metres or so of seawater is on average westward away from the coast because of Ekman transport (see Figure 10). This mass of displaced seawater can only be replaced by seawater from below, which is drawn up to the surface. If the upwelling is coming from water beneath the depth of the thermocline, then this replacement seawater is cold and nutrient rich. Likewise, in the northern hemisphere, when a steady wind blows roughly from the north along the western edge of a continent, the surface layer of seawater moves on average westward away from the shore, creating coastal upwelling conditions. Since coastal upwelling is dependent on favourable winds, it tends to be a seasonal or intermittent phenomenon and the strength of upwelling will depend on the direction and strength of the winds.

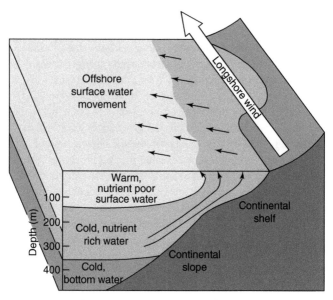

**10. Depiction of coastal upwelling along the west coast of a continent in the southern hemisphere bringing nutrient-rich seawater to the surface.**

Important coastal upwelling zones around the world include the coasts of California, Oregon, north-west Africa, and western India in the northern hemisphere; and the coasts of Chile, Peru, and south-west Africa in the southern hemisphere. These regions are among the most productive marine ecosystems on the planet. When upwelling is occurring, the cold, nutrient-rich seawater stimulates immense blooms of phytoplankton, mainly the larger types such as diatoms and large species of dinoflagellates.

## El Niño Southern Oscillation

The upwelling off the west coast of South America supports one of the most productive fisheries on the planet—the Peruvian anchovy fishery. Peruvian fishers have long been aware of a phenomenon they termed El Niño (the Child) in honour of the Christ child, because it often occurred around Christmas. During El Niños the surface waters become unusually warm and fish and seabirds die, the anchovy fishery diminishes or collapses, and whales, dolphins, and seals disappear.

We now know that this is part of an important and poorly understood global phenomenon termed the El Niño Southern Oscillation (ENSO). ENSOs are associated with periodic reversals of atmospheric pressure in the Pacific Ocean. Normally, there is a persistent high-pressure system in the eastern Pacific Ocean, and a persistent low-pressure system in the western Pacific Ocean. Under these conditions the Pacific trade winds blow strongly from east to west, moving surface seawater towards the west and creating a coastal upwelling system that supports the Peruvian anchovy fishery. During an El Niño event, though, this pressure system reverses for an unknown reason. This causes the trade winds to diminish, allowing warm water from the western Pacific Ocean to move eastwards and accumulate along the coast of South America, depressing the upwelling system that supports the anchovy fishery. Without the anchovy, seabirds die and marine mammals that normally feed on anchovy leave their usual feeding grounds. El Niños occur about every two to seven years and normally end in less than a year, although severe El Niños can persist for several years.

More than twenty El Niño events have been recorded since the early 1950s and there is evidence that the intensity of these events has increased in recent years. During this period, the most severe El Niños occurred in 1982–3, 1997–8, 2014–16, and 2018–19. The effects of strong El Niños are felt globally. California and western South America can experience unusually heavy rains, while parts of Australia, Indonesia, and Africa can experience severe drought. The flooding, crop failures, landslides, and other events associated with these El Niño weather patterns are very costly, causing huge damage and many deaths.

# Moving energy through the marine food web

As discussed in this chapter, phytoplanktonic microbes suspended in the photic zone are responsible for almost all the primary production in the oceans. But what are the various pathways through which this food energy moves through the marine food web and with what efficiencies? Over the past four decades marine biologists have made significant advances in answering these questions.

Up until the 1970s it was thought that most of the primary production in the oceans moved through a simple food chain comprising a small number of 'trophic levels'—the level in a food web from which a species derives its food. In this 'classic' view, the primary producers at the first trophic level are mainly large diatoms and dinoflagellates which are eaten by large zooplankton, mainly copepods, at the second trophic level (see Figure 11(a)). These in turn are eaten by small fish at the third trophic level, which are then food for top predators such as squid, large fish, and marine mammals at the fourth trophic level. In our human-dominated world of today, many of these marine top predators are harvested by the ultimate top predator—humans at the fifth trophic level (see Chapter 8).

When marine biologists began to understand the importance of microbes in the economy of the oceans, it became clear that this classic food chain concept was in many cases too simple and incomplete a picture of how energy moves through the marine ecosystem. Consequently, new models of food web structure have emerged which are still being studied and refined by marine biologists.

It is now known that microscopic autotrophic bacteria produce a significant amount of the primary production in the oceans, particularly in nutrient poor oceanic waters. Their cells are too

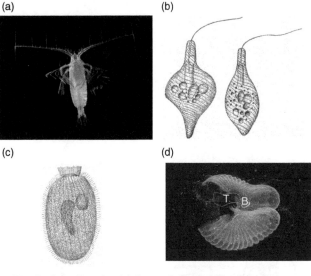

(a) (b)

(c) (d)

11. Zooplankton diversity: (a) Copepod; (b) Flagellated protists; (c) Ciliate protist; (d) Giant larvacean. B = body of animal, T = tail of animal © 2002 MBARI.

small to be captured and eaten by the large zooplankton of the classic food chain, but small flagellated protists (see Figure 11(b)) can consume them. These are then eaten by ciliated protists (see Figure 11(c)) which are large enough to be eaten by larger zooplankton, such as copepods, thus linking into the classic food chain (see Figure 12).

It is also now known that much of the phytoplankton production—up to 50 per cent under certain conditions—is not transferred to the next trophic level through direct consumption but is 'leaked' into the surrounding seawater as dissolved organic matter (DOM). Some of this leakage results from passive loss of organic molecules across the membrane of phytoplankton cells. But phytoplankton also actively exude organic molecules into the seawater for reasons not yet entirely understood.

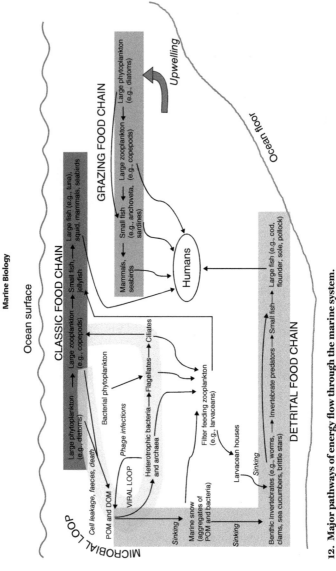

12. **Major pathways of energy flow through the marine system.**

Furthermore, DOM and small particles of organic matter (POM) are lost when phytoplankton cells are broken up during feeding by predators and following the natural death of the cells. Copepods and other zooplankton also release large amounts of POM into the oceans in the form of faeces and when they die and disintegrate. It is also recognized that viral phage infections play a very important role in the loss of DOM and POM from marine microbes. It has been estimated that viruses kill about 20 to 40 per cent of marine bacteria daily—a remarkable infection rate—and large amounts of DOM and POM are released as the host cells die.

The loss of organic matter from all these sources creates an enormous pool of non-living food energy in the oceans (see Figure 12). The total size of this pool, expressed in carbon, is estimated to be about 1,000 billion tonnes, which dwarfs the roughly 6 billion tonnes of carbon existing within all the living organisms of the oceans. Much of this is DOM in the form of carbohydrates, amino acids, proteins, and lipids.

Heterotrophic bacteria and archaea absorb DOM directly from the seawater and use it as an energy source. They also colonize POM floating in the seawater and extract food energy from it. Many of these microbes are quickly killed by viruses and thus re-release DOM and POM back into the seawater. This rapid cycling of energy back and forth between the pool of DOM and POM and bacteria and archaea, mediated by phages, is often referred to as the 'viral loop' (see Figure 12). Those bacteria and archaea that escape viral infection are consumed by flagellates and ciliates. These protists then link into the classic food chain discussed earlier when eaten by larger multicellular zooplankton, such as copepods (see Figure 12).

These microbial pathways of energy flow are often referred to as the 'microbial loop' because when depicted diagrammatically they form a 'loop' connected to the classic food chain (see Figure 12). The flow of energy through the six trophic levels in the combined microbial loop and classic food chain—from bacteria and archaea through to top predators like squid, tuna, and seabirds—involves five transfers of energy. Only about 10 per cent of the energy available in each trophic level can be converted into biomass at the next trophic level. This is because much of the food energy consumed by the organisms in each trophic level is lost as faeces and expended on basic metabolic processes and searching for food. Furthermore, not all the available food at each trophic level can be located and consumed by organisms at the next level, so there is 'wastage'. Therefore, only a tiny fraction of the original primary production—about 0.001 per cent—ends up as flesh in the top predators at the sixth trophic level. Despite its length and comparative inefficiency, this food chain is an important nutritional link between the huge amount of non-living organic matter in the oceans and large marine animals, which are exploited by humans. It is especially important in the extensive open-ocean regions of the Global Ocean where most of the primary production is from autotrophic bacteria which are too small to enter the classic food chain.

Some of the POM created at the surface of the oceans is 'exported' into deeper waters as it sinks slowly through the water column (see Figure 12). On the way down, smaller organic particles aggregate to form larger particles which are colonized by heterotrophic bacteria and become 'marine snow'—a shower of small whitish flakes of detritus several millimetres in size that can be observed in the lights from a submersible or remotely operated underwater vehicle. Bacteria extract a considerable amount of the energy in marine snow before it reaches the ocean floor, particularly in very deep water over the abyssal plains. Nonetheless, marine snow represents an important food source for marine animals living in the deeper waters of the oceans and on the ocean floor.

Some types of zooplankton have evolved fascinating ways of using marine snow as an energy source. For example, small tadpole-like animals known as larvaceans construct ornate 'houses' out of thin films of mucus. In some species, these houses can be up to a metre across (see Figure 11(d)). The beat of the larvacean's tail creates a current through the house which filters organic particles, which are then passed to the animal's mouth. The larvacean discards its house every few hours once it becomes clogged and quickly secretes a new one. These discarded houses, along with all the particles trapped on them, collapse and sink quickly to the ocean floor before bacteria can extract all the energy from them, thereby increasing the input of food to animals on the ocean floor.

The detrital material that reaches the ocean floor creates the base of a benthic food chain (see Figure 12). The detritus is eaten by animals such as worms, clams, and brittle stars, which are then eaten by predatory animals such as sea stars, crabs, and small bottom-dwelling fish. These animals are then eaten by larger predators including fish of commercial value, such as flounder, sole, and pollock (see Chapter 8).

Grazing food chains, similar in structure to the classic food chain, operate in upwelling zones (see Figure 12). In these systems, diatoms are generally the main primary producers and occur in huge, dense blooms. Massive schools of small, commercially valuable filter-feeding fish, such as anchoveta and sardines, graze directly and efficiently on the diatoms and on copepods feeding on the diatoms. These small fish are then eaten by large fish, marine mammals, and seabirds. The grazing food chain involves only one or two transfers of energy between phytoplankton at the first trophic level and commercially valuable fish at the second or third trophic levels (see Figure 12). Furthermore, the efficiency of energy transfer between trophic levels in these simple food chains involving dense aggregations of organisms can be as high as 20 per cent. Upwelling systems are thus very efficient and productive and support large commercial fisheries (see Chapter 8).

# Trophic structure of marine and terrestrial systems—a study in contrasts

The total primary productivity of the Global Ocean is about 50 billion tonnes of carbon per year. In comparison, total terrestrial primary productivity is around 52 billion tonnes per year. Thus, the total primary productivity of the planet is a little over 100 billion tonnes of carbon per year, of which about half comes from the oceans and half from land. But the ocean and land systems generate primary production in fundamentally different ways.

As discussed in this chapter, the primary producers in the oceans are invisible, single-celled, microscopic phytoplankton comprising bacteria and protists. In contrast, the primary producers on land are large, highly visible plants comprising trees, shrubs, grasses, and, now that we have entered the human dominated Anthropocene epoch, food crops. The total biomass, or standing crop, of these very different types of producers is also starkly different. The total biomass of primary producers in the oceans, expressed as the weight of carbon in their living tissues, is about 1 billion tonnes, whereas the total biomass of primary producers on land is several orders of magnitude higher—about 450 billion tonnes of carbon. Thus, marine primary producers, although incredibly numerous and occupying a much larger surface and volume than the terrestrial environment, account for only about 0.2 per cent of the total biomass of primary producers on the planet. Yet, remarkably, marine primary producers generate as much primary production as all the producers on land on an annual basis. This is made possible because marine phytoplankton have extremely high metabolic rates and are highly efficient photosynthetic engines, which allows them to grow and divide very rapidly. The result is that phytoplankton in the Global Ocean are turning over their entire biomass about every seven days on average. In contrast turnover rates of terrestrial primary producers are on the order of years on average.

**13. Distribution of biomass between producers and consumers in the terrestrial and marine environments. Numbers are billions of tonnes of carbon.**

The trophic structures of the marine and terrestrial systems are also fundamentally different. On land, the approximately 450 billion tonnes of carbon of primary producer biomass supports a total of roughly 20 billion tonnes of consumers in all the higher trophic levels. This results in what is generally considered to be a typical standing biomass pyramid, where the biomass of primary producers is much larger than that of the consumers. In contrast, in the oceans about 1 billion tonnes of primary producers support a total of about 5 billion tonnes of consumers, resulting in an inverted biomass pyramid (see Figure 13). In an ecological food web, primary productivity must always be greater than total consumer productivity. The 'atypical' standing biomass distribution in the marine environment can be maintained because, as discussed earlier, the microbial primary producers of the ocean have a very rapid turnover, so their resulting productivity is necessarily higher than the total consumer productivity.

# Chapter 3
# Life in the coastal ocean

The coastal regions of the Global Ocean comprise a narrow strip of ocean extending from the shoreline to the edge of the continental shelf. Although this coastal ocean environment is comparatively small, accounting for only about 7 per cent of the area of the Global Ocean, it is of huge importance to human society. Currently, roughly 44 per cent of the human population, or 3.4 billion people, crowd along the coast or live within 150 kilometres of a coast. This number is increasing rapidly as more people migrate to urban centres near coastal regions. It has been forecast that by the end of the century thirteen out of fifteen of the world's largest cities will be located on or near a coast.

This close relationship with human society means that the coastal ocean is heavily impacted. It is the receiving environment for many of the by-products of human activities, such as industrial and agricultural pollutants, human and animal sewage, plastic waste, and oil spills. The coastal ocean is heavily fished, providing much of the wild-caught seafood that humans obtain from the oceans. It is also the place where most aquaculture operations are sited, providing additional seafood for human consumption, but also leaving behind aquaculture waste products, such as unconsumed feed and faecal material, which accumulate and decompose on the ocean floor. Furthermore, the coastal ocean is heavily exploited for other resources such as oil and gas.

Because of its proximity to land, and its comparative shallowness, the coastal ocean is comparatively easily studied by means of scuba diving, small research vessels, moored instruments, and, more recently, networks of thousands of Internet-connected sensors gathering real-time continuous data on ocean physics, chemistry, and biology. Thus, much information is being gathered about the biology of coastal marine life and how coastal ecosystems function. If put to good use, this knowledge will help us to better manage the various habitats comprising the coastal environment in the face of a rapidly growing human presence.

## Kelp bed habitats

Kelp beds are beautiful and important marine habitats found on rocky bottoms in shallow waters close to shore. They are found in cool, temperate coastal regions where ocean temperatures normally do not exceed 20°C. They thrive particularly well in the cool, nutrient-rich waters of coastal upwelling zones (see Figure 14).

Kelp beds consist of dense aggregations of brown seaweeds, or kelps. Whereas most photosynthetic organisms in the oceans are microscopic, single-celled, and planktonic, kelps are large, multicellular algal forms that are attached firmly to the bottom by root-like structures known as holdfasts (see Figure 15). The holdfast is an attachment structure, not a root system, and kelps do not possess a specialized vascular system for transporting food and nutrients through its frond. Instead, the kelp absorbs nutrients, water, and carbon dioxide directly from the surrounding seawater.

Kelps begin life as microscopic spores which settle to the ocean floor and grow into microscopic male or female plants, called gametophytes. The gametophyte is the sexual stage of the life cycle. The male gametophytes produce sperm, which are released into the ocean where they fertilize eggs present on the surface of the female gametophytes. The fertilized eggs grow into the large fronds, or sporophytes, that we generally associate with kelps.

Life in the coastal ocean

49

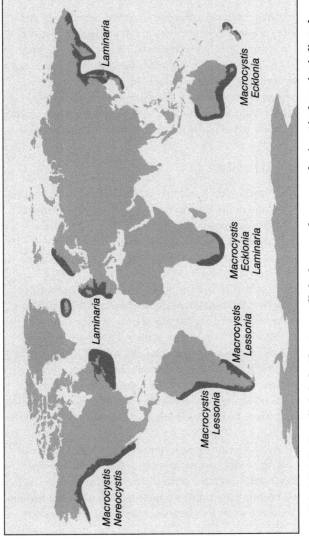

14. Global distribution of kelp beds. The genera of kelp shown on the map are dominant in the region indicated.

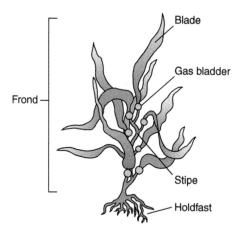

Blade

Gas bladder

Frond

Stipe

Holdfast

**15. The structure of a kelp frond.**

The sporophytes then release many new spores into the ocean to start the process over again. The life cycle takes around one to two years to complete.

Kelp beds are very productive. Under favourable conditions kelps grow very quickly, and some species reach a very large size. For example, the giant kelp, *Macrocystis*, which is abundant along the coasts of many parts of the world (see Figure 14), can grow at rates of more than 30 centimetres per day and reach lengths of more than 30 metres in less than a year. Giant kelps form vast underwater 'forests' with dense surface canopies that are held aloft in the seawater column with the assistance of gas-filled floats called pneumatocysts.

Kelp beds provide the foundation for a very lush, diverse marine habitat. Many types of organisms live attached to the surface of the kelps themselves, or buried within the interstices of the holdfasts, and the kelps provide shelter and food for many species of invertebrates and fish, including commercially harvested species.

Sea urchins are the main grazers of kelps, although they normally remove very little, perhaps 10 per cent or so, of the living kelp biomass directly, unless present in very large numbers. Much of the energy in the kelps enters the kelp bed community in the form of dead plant material, which is consumed by a range of scavengers, including sea urchins and, where present, abalone. The sea urchins are fed on by a range of predators including species of sea stars, snails, octopus, spiny lobsters, crabs, and fish. In the North Pacific Ocean, sea otters also feed on sea urchins.

Longer-term studies of kelp habitats have shown that they are susceptible to overgrazing by sea urchins. When this happens, luxuriant, productive, and biologically diverse kelp communities are converted into drab, impoverished 'urchin barren grounds' (see Figure 16(a) and (b)). Although these 'phase shifts' occur naturally on a limited scale, they have recently become a more frequent and global problem. Rapid, widespread, and catastrophic collapse of kelp beds has occurred worldwide, including in the

(a)                                   (b)

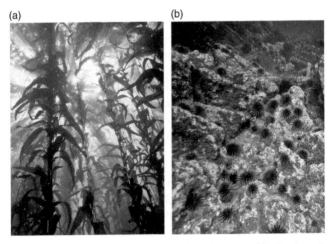

16. Kelp bed collapse: (a) Typical kelp bed; (b) A sea-urchin-dominated barren ground that has replaced a kelp bed.

north-east Pacific Ocean in the 1960s–1970s, the north-west Atlantic Ocean in the 1970s–1980s, along the coast of Norway in the 1970s, and, most recently, along the east coast of Tasmania in the 2000s.

Marine biologists have made good progress over the last three decades in unravelling the various factors that can stress a kelp bed habitat to the point where it becomes susceptible to a rapid transition into an urchin barren. Both physical and biological stressors have been identified that often interact to exert an additive effect. Most are human-derived or amplified by human activities.

Physical stress on kelp beds can take the form of higher than normal ocean temperatures or unusually rough sea conditions. For example, periods of unusually warm seawater on the Pacific coast of North America, brought about by a reduction or cessation of upwelling of deep cool water, can stress kelps to the point where they are vulnerable to urchin overgrazing, particularly if urchin numbers are higher than normal for other reasons. Also, large storm-generated waves associated with severe El Niño events can shred kelps to pieces and tear many of them from their holdfasts. The sea urchins in the area can then take over, grazing down any young kelps attempting to recolonize the area and hindering the re-establishment of a kelp habitat. Human-induced climate change is increasing the frequency, duration, and magnitude of such physical stresses on kelp beds, making them more susceptible to overgrazing and collapse.

Decreases in the abundance of a key sea urchin predator as a result of human interference is a well-studied contributor to kelp bed collapse. A classic example is the predator–prey interaction between sea otters and sea urchins. By feeding on sea urchins, sea otters keep the intensity of grazing on kelps below a critical level which helps maintain a stable kelp bed community. But if sea otters are largely missing from the system sea urchin numbers can increase greatly, putting greater grazing pressure on kelp beds.

Once a critical density of sea urchins is reached, kelp beds rapidly collapse into urchin barren grounds. For this reason, sea otters are often referred to as a keystone species, which is one that may not be present in large numbers but plays a fundamental role in shaping and maintaining the structure of a marine community.

Sea otters were historically an abundant member of kelp communities in the coastal ocean along the rim of the northern Pacific Ocean from northern Japan to Baja California. Their total natural population was estimated to be in the range of 150,000 to 300,000.

There is evidence from the study of middens that over 2,000 years ago aboriginal Aleut fishers in the Aleutian Islands hunted sea otters heavily enough to cause a localized shift in the structure of the coastal marine community, from kelp dominated to sea urchin dominated. Commencing in the 1700s, sea otters began to be harvested on a widespread industrial scale by fur traders for their dense pelts, and by the early 1900s only several thousand sea otters remained throughout their entire range, inhabiting just a few isolated coastal refugia. This resulted in the creation of extensive urchin barren grounds in places like the Aleutian Islands and off the coasts of Alaska, Canada, and elsewhere. Commercial hunting of sea otters was mostly banned from 1911 and conservation efforts commenced, with sea otters from surviving populations being reintroduced into certain areas. As a result, sea otter populations have rebounded to over 100,000 animals occupying two-thirds of their former range, and a shift back to a kelp-dominated habitat has been observed in many areas where sea otters are now abundant.

The recent decimation of kelp beds off eastern Tasmania is another example of a human-derived impact. Here, ocean warming has allowed large numbers of the sea urchin *Centrostephanus rodgersii* to spread from mainland Australia poleward into Tasmania, causing extensive overgrazing of kelp and the collapse of over 95 per cent of the eastern Tasmanian kelp beds.

Once created, urchin barrens can persist for many decades. Part of the reason is that once the sea urchins have consumed all the kelps, their numbers persist because they can survive by feeding on other sources of food including encrusting algae growing on bare rock surfaces and on small invertebrates. It often takes another major event to trigger a shift back to a kelp habitat. This can take the form of a massive mortality of the sea urchin population, which reduces grazing pressure on young kelp plants, and clears the way for a kelp habitat to be re-established. Such urchin mass mortality can be the result of an urchin disease and, ironically, anthropogenic ocean warming is likely to increase the frequency of such disease events and assist kelp recovery. This is because periods of unusually warm seawater can stress sea urchins and make them more susceptible to pathogens.

There is still much to learn about the complex dynamics of kelp bed collapse and recovery and further research will be crucial for the future management of kelp bed habitats and the retention of their ecological and social benefits, particularly in an era of rapidly increasing human-derived impacts on the oceans.

## Seagrass meadows

Seagrasses are the basis of another important type of marine habitat that is widespread on sandy and mud bottoms in shallow waters from the tropics to the Arctic, with Antarctica being the only continent without seagrasses. Unlike kelps, seagrasses are flowering plants that have become adapted to live completely submerged in seawater. They evolved originally on land and are the only group of flowering plants that have recolonized the marine environment.

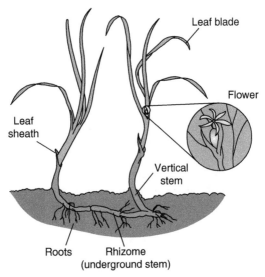

**17. Structure of seagrass.**

Seagrasses have a true root and vascular system and absorb their nutrients from the sediment on which they live (see Figure 17). Their leaves are long, thin, flexible blades, generally about 10–50 centimetres in length, but up to several metres in some species. Seagrasses colonize the seabed by sending out underground stems, or rhizomes, which give rise to new plants. They can also reproduce sexually, with currents carrying pollen from one flower to another, and with the seeds also being dispersed by currents.

Seagrasses are very productive. The blades can grow as much as a centimetre per day and are continually being shed and replaced. Seagrasses can thus form vast meadows covering tens of thousands of hectares and comprising thousands of leaves per square metre of seabed. Seagrass meadows are most abundant in water less than about 10 metres deep where light levels are high, but they can be found to depths of over 40 metres in very clear seawater. In temperate regions, *Zostera*, or eelgrass, is a widespread type of seagrass; while *Thalassia*, or turtle grass, is common in tropical regions.

Seagrasses form the basis of a diverse community. Many kinds of clinging and encrusting organisms colonize the surface of the seagrass blades, while various types of burrowing animals, such as clams and worms, are found among the roots. These organisms provide a rich source of food and shelter for fish and other animals, including commercially important species.

In temperate regions there are few animals that can graze directly on the tough seagrass blades, except for some birds, such as swans, geese, and ducks. In tropical waters, large grazers such as manatees, dugongs, and green turtles feed directly on living seagrass leaves. The dead leaves and roots of seagrasses are broken down by bacteria and fungi and fed upon by a range of detritus feeders such as worms, crabs, brittle stars, and sea cucumbers.

The green turtle, *Chelonia mydas*, is intimately associated with seagrass meadows. Green turtles are key grazers of seagrass, hence the common name, turtle grass, for some tropical species of seagrass. An adult green turtle can eat about 2 kg of seagrass a day.

Green turtles mate at sea and the females return to the same beaches where they hatched to lay their own eggs. A female digs a pit on a dry area of the beach into which she lays her eggs. She then covers the eggs with sand and returns to the ocean. A female may return several times to the same beach during the breeding season at intervals of several weeks to lay successive clutches of eggs. The eggs hatch after a couple of months, generally at night, and the hatchlings scuttle quickly to the ocean. Here they remain in coastal waters near where they hatched to feed and grow. When they mature, the turtles migrate to new seagrass feeding grounds where they remain until they are ready to return to their nesting beaches to reproduce. These migrations can cover vast distances. For instance, one population of green turtles feeds on seagrasses along the Brazilian coast but migrates over 2,300 kilometres to nesting beaches on tiny Ascension Island in the middle of the South Atlantic Ocean to breed. The turtles use the angle of the sun, wave direction, and smell to navigate their way to the island.

Worldwide, green turtles have been heavily hunted for their meat, and their eggs are harvested from their nesting beaches for food. Consequently, green turtle numbers have declined drastically and they are a globally endangered species. It has been estimated that in pre-Columbian times there were over 90 million green turtles inhabiting the Caribbean Sea alone. In their diaries, explorers describe green turtles in the Caribbean in terms of 'infinite numbers' and 'inexhaustible supplies'. Since then there have been staggering losses, with numbers now down to about 300,000, or 0.33 per cent of their historic numbers. Since their numbers are now so low, their effect on seagrass communities is currently small, but in the past, green turtles would have been extraordinarily abundant grazers of seagrasses. They would have kept seagrass meadows cropped, like putting greens, and thinned out the beds, thereby reducing direct competition among different species of seagrasses and allowing several species to coexist. Without green turtles in the system, seagrass beds are now generally overgrown monocultures of one dominant species.

Manatees and dugongs are large marine mammals that also graze on seagrasses. Dugongs live in the warm waters of the Indian and Pacific Oceans, whereas manatees are found in the Caribbean and Gulf of Mexico, and off the coast of West Africa. Dugongs in Australia have been shown to engineer seagrass beds like green turtles. Herds of grazing dugongs thin out the beds, creating space for more species of seagrass to coexist and for younger, more nutritious, plants to regenerate. This has been likened to dugongs 'cultivating' the seagrass meadows in a way that provides them with improved nutrition.

Seagrasses are of considerable importance to human society. Because they are effective at trapping and binding sediment, they stabilize the ocean bottom and help protect coastlines from erosion. Seagrass beds also act as nursery grounds and foraging areas for many commercially important species of fish, as well as commercially important invertebrates such as clams, crabs, shrimp, and oysters.

It is therefore of great concern that seagrass meadows are in serious decline globally. A comprehensive global assessment has shown that 29 per cent of known seagrass meadows disappeared between 1879 and 2009. The rate of loss was about 1 per cent per year prior to 1940, but this has accelerated to about 7 per cent per year since 1990. Much of this is the result of sewage discharges, coastal development, and seabed dredging, activities which release excessive amounts of nutrients and sediment into coastal waters which reduce water clarity and smother seagrasses. Seagrass meadows are also routinely torn up by boat propellers and anchors. Furthermore, seagrasses stressed by pollution or elevated ocean temperatures are more susceptible to disease.

It is obvious that humans need to value seagrass habitats more highly and take the necessary steps to preserve them. This means protecting important seagrass habitats from coastal development and educating recreational boaters to not drive across shallow stretches of seagrass, and to anchor away from seagrass meadows.

## Soft-bottom communities

Vast expanses of the coastal regions of the Global Ocean consist of sand or mud bottoms. These habitats are strongly influenced by currents and wave action, especially in the shallower depths. Sandy bottoms are present where there are reasonably strong currents that carry away the fine, easily suspended mud particles, leaving behind the coarser sand grains. Muddy bottoms occur in areas with little current and tend to be rich in fine particles which settle out and collect in such areas. In either case, there is almost a complete lack of vegetation, and animal life dominates these habitats.

Most of the animals living in soft-bottom habitats are found buried in the sediment and are referred to as infaunal organisms. For the most part these comprise various species of burrowing clams and worms. Many of these are suspension feeders which filter plankton and organic particles from the overlying seawater. Suspension-feeding clams extend long siphons from their burrows up to the surface of the sediment and pump in this suspended food material. The suspension-feeding worms often dwell in tubes from which they extend a set of tentacles used to filter out suspended particles. On the other hand, some of the infaunal clams and worms are deposit feeders, obtaining nutrition by directly consuming sediment and digesting the organic material and bacteria present in the sediment as it passes through their guts.

Not all members of the soft-bottom communities are burrowers. Some, referred to as epifaunal organisms, live on the surface of the sediment and include brittle stars, sea urchins, sea stars, sea cucumbers, sand dollars, snails, crabs, and shrimp. Many of these surface dwellers feed by either picking up organic particles lying on the surface of the sediment or, like infaunal deposit feeders, engulfing sediment and digesting the organic particles it contains. Others, such as snails, crabs, and sea stars, are predators, feeding on other members of the community.

The infaunal and epifaunal members of soft-bottom communities support a range of bottom-feeding predatory fish that live on or near the ocean floor. These include skates and rays and many types of commercially important species such as haddock, pollock, hake, and cod, as well as flatfish such as flounder, halibut, and sole. These bottom-dwelling fish, collectively known as demersal fish or groundfish, are generally caught using bottom trawls, and play a vital role in feeding the human population.

## Coastal dead zones

Many coastal marine systems are greatly stressed and altered by the release of excess nutrients into coastal waters because of human activities, a process referred to as eutrophication. The two main culprits are compounds containing nitrogen and phosphorus.

The nitrogen cycle is one of the most altered nutrient cycles on the planet due to human activity. The use of nitrogen-based fertilizers on land to maintain soil fertility and boost crop yields is a major source of nitrogen pollution. This is because a portion of the nitrogen compounds spread on agricultural land is not incorporated into crops but is leached by rain into streams and rivers and, hence, into the oceans. Much of the nitrogen leaching from cropland is in the form of nitrate.

Other sources of nitrogen pollution are sewage discharges of human and animal wastes. Nitrogen compounds are also produced during the burning of fossil fuels, wood, and crop waste; these end up in rivers and the oceans as nitric acid in acid rain.

The extensive use of phosphate-based fertilizers is a major source of phosphorus pollution. Some of the phosphate is washed off into streams and rivers, mostly adhered to soil particles; from there it finds its way into the oceans. Human and animal sewage waste is another source of phosphorus compounds, as is deforestation. In the latter case, some of the soil exposed following the removal of trees, as well as the ash from burnt trees, both of which contain phosphorus, is washed into rivers. It is estimated that phosphorus concentrations in many rivers are now on average twice natural levels, and much of this ends up in the oceans.

Eutrophication of the oceans through excess nitrogen and phosphorus stimulates massive blooms of phytoplankton, particularly in coastal waters. As this mass of primary producers dies, it decays and consumes oxygen in the seawater. If dissolved oxygen levels are reduced below what is required to sustain most of the marine life in the area, about 2 millitres per litre of seawater, then temporary or permanent marine 'dead zones' are the result. Loss of biodiversity, high fish mortality, and the collapse of local fisheries are all associated with dead zones.

The number of marine dead zones in the Global Ocean has roughly doubled every decade since the 1960s and now sits at over 500, up from fewer than 50 in 1950. Not surprisingly, dead zones are common in areas subject to run-off from intensive agriculture. The Gulf of Mexico dead zone is a seasonal feature that results from the huge amount of nutrients discharged in the spring into the Gulf from the Mississippi River, which drains a large area of intensively farmed agricultural land. This dead zone routinely covers an area of about 15,000 to 18,000 square kilometres, although in 2017 it was close to 23,000 square kilometres. Apart from the overall destruction of a natural marine system, this dead zone impacts negatively on commercial and recreational shrimp and oyster fisheries in the Gulf. Other dead zones include parts of the Baltic Sea, the northern Adriatic Sea, Chesapeake Bay, and numerous sites in coastal areas of east Asia. The Baltic Sea dead zone is one of the largest on the planet and now regularly covers an area of about 70,000 square kilometres—roughly the size of Ireland.

Dead zones are harmful features in the coastal oceans which are rapidly increasing in number and size. This trend can be halted and ultimately reversed by widespread implementation of precision agricultural practices that optimize and reduce fertilizer use and investment in improved sewage management systems.

# Harmful phytoplankton blooms

Blooms of toxic phytoplankton occur frequently in coastal waters. These are often referred to as 'harmful algal blooms' (HABs). Phytoplankton cell densities in these blooms can be high enough to discolour the ocean, sometimes to a reddish hue, hence their common name 'red tides'.

HABs are created by a small number of phytoplankton species, often dinoflagellates, which produce a range of potent toxins that can be excreted into the seawater. These toxins are then transferred through the food web, accumulating first in zooplankton that feed on the toxic phytoplankton and in animals such as clams, mussels, scallops, and oysters that filter-feed on the toxic plankton. The toxins can then be transferred further up the food chain into fish, marine birds, and marine mammals. All these organisms can be affected to a greater or lesser extent through the consumption of these toxins, sometimes resulting in massive mortalities of fish and other marine life, including seabirds and marine mammals, and causing the closure of local fisheries.

Humans feeding on contaminated shellfish and fish can experience neurological symptoms, such as tingling of the fingers and muscular paralysis, as well as respiratory problems and gastrointestinal symptoms such as diarrhoea, vomiting, and abdominal cramps. Such symptoms can be very severe and sometimes fatal. Toxic phytoplankton can sometimes become suspended in the air by wave action to form a coastal spray which, when inhaled by people, can cause asthma-like symptoms.

Occasional HAB events are a natural phenomenon and have probably always been a feature of coastal waters. Early explorers of the 17th and 18th centuries described occurrences of discoloured water and noted that indigenous coastal peoples would avoid harvesting shellfish at certain times of year in certain places for fear of being poisoned. But in recent decades coastal regions throughout the world are experiencing HAB events with much greater frequency, persistency, and geographic extent, and involving a greater number of phytoplankton species.

It is not entirely clear why this is occurring. It may in part be explained by improved recognition and reporting of toxic bloom and seafood poisoning events, but there is almost certainly a link to increased eutrophication of coastal waters that promotes phytoplankton blooms. Also, toxic phytoplankton species are being routinely transferred from one port to another around the world in the ballast water of ships. Ballast water is seawater pumped into the ballast tanks and cargo holds of ships to give them better stability when on a voyage. Ballast water is usually taken on when a ship has delivered cargo to a port and is leaving with less cargo or no cargo. Millions of litres of seawater are taken on at a time and then often transported and released at the next port where the ship picks up more cargo. This is a likely mechanism for expanding the range of toxic phytoplankton species. Some species can produce cysts—a long-lived resting phase of their life cycle—that can lie dormant in the bottom sediments for many years until conditions are favourable for growth, and a HAB event is triggered.

## Biological invasions

Many other kinds of marine organisms in coastal waters besides toxic phytoplankton can be pumped into the ballast tanks of ships. When a ship is in shallow water it can also pump in sediments and any associated bottom-dwelling organisms. When the ballast water is next released these organisms may also be released. In this way non-native, or exotic, invaders are introduced into areas where they would never normally be found without human involvement. Roughly ten billion tonnes of ballast water are transferred globally each year and thousands of marine species are carried around the world in ballast water every day.

Ships also move marine organisms long distances in other ways—boring organisms, such as ship worms, colonize the hulls of wooden ships, and attached organisms like barnacles and seaweeds foul the hulls of ships. These organisms can release their free-living larval stages in foreign ports which then settle to the bottom, allowing the organisms to colonize the new site.

Typically, very few of these foreign invaders will survive in their new surroundings. Nonetheless, some encounter conditions that allow them to become well established and sometimes overwhelm the natural marine community in the area. This may be because the invader lacks the natural predators, pathogens, or parasites in its new location that would normally keep its numbers in check. Or it may encounter an unusually abundant food supply or is able to outcompete native species for available food and habitat space. Examples of introduced marine organisms are legion and include various species of seaweeds, jellyfish, sponges, worms, crabs, barnacles, sea stars, clams, mussels, oysters, snails, fish, and many others.

The introduction of a jellyfish-like animal, the comb jelly, *Mnemiopsis leidyi*, from the coastal waters of North America into the Black Sea in ballast water in the 1980s illustrates very well the devastation that can be caused by marine invaders. This animal quickly multiplied to plague numbers in the predator-free environment of the Black Sea, voraciously consuming the natural zooplankton in the sea, including the eggs and juvenile stages of fish. Fish stocks collapsed by the early 1990s, causing great economic loss to the region, and dolphins, which fed on these fish, disappeared. Interestingly, it took the invasion of another exotic species of comb jelly, *Beroe ovata*, also in ballast water, to alleviate this ecological disaster. Around 1997, this species began to thrive in the Black Sea, feeding heavily on the first foreign invader, causing a steep reduction in its numbers. The *Beroe ovata* population then collapsed as it exhausted its food supply. Since then, fish stocks have begun to recover, and dolphins have returned.

The introduction of the Japanese sea star (*Asterias amurensis*) into Australia is another good example of the major impact that an exotic species can have when it invades a new habitat. This sea star is native to the coastal waters of Japan, northern China, North and South Korea, and Russia, but sometime in the 1980s it was introduced into Tasmania, probably as larvae in ballast water, or as juveniles clinging to the hulls of ships arriving from the North Pacific.

The population of this sea star exploded in its new location and by the mid-1990s had reached extraordinary densities in some places. For example, in the Derwent Estuary of Tasmania there are an estimated thirty million individuals at densities up to 10 per $m^2$. This sea star is a voracious predator feeding on just about anything in its path, including shellfish, crabs, sea urchins, sea squirts, and other sea stars, and turning the bottom into a virtual monoculture of foreign sea stars. They also pose a threat to aquaculture operations in the area with the potential to decimate mussel, oyster, and scallop farms.

Once well established, these foreign invaders are impossible to eradicate. Attempts have been made to control the spread of the Japanese sea star by recruiting divers to remove the animals by hand, and by trapping or dredging the sea stars. They have also been commercially harvested and converted into fertilizer. But none of this has had much success in restoring the natural marine system of the area. Efforts are now focused on limiting the spread of the species through an education campaign that encourages reporting of local sightings, which is followed up by a rapid eradication programme.

International efforts are being made to limit the spread of exotic marine species in ballast water. Ships are meant to empty and then recharge their ballast tanks in the open ocean before arriving at a port. The reasoning behind this is that ballast water hitch-hikers taken up in port will be released into the open ocean where they cannot survive, and that the planktonic organisms taken up in the open ocean will be released into the coastal waters of the next port where conditions will not be suitable for survival. Many countries also require that ballast water is treated with a shipboard system to remove and/or kill organisms before discharge in a port. This can involve treatments using filtration, heat, or disinfectants, often in combination.

## Plastic debris

Over the past sixty years, plastic materials, which are derived from oil and gas, have become a ubiquitous and indispensable product of human society, and a major source of human waste. In 1967 about 2 million tonnes of plastics were produced per year; it is currently about 380 million tonnes per year. It has been calculated that humans have produced a total of about 8.3 billion tonnes of plastic to date, more than three-quarters of which has been discarded and ended up in landfills or accumulating in the wider environment.

Not surprisingly, huge amounts of this waste plastic have entered the marine environment—about 10 million tonnes per year currently. Around 80 per cent of marine plastic debris comes from the land, either directly or via rivers. Ten rivers, two in Africa and the rest in Asia, discharge about 90 per cent of all marine plastic debris. Ships and boats are another source of marine plastic pollution. A great variety of plastic trash is dumped overboard from commercial and recreational vessels, and fishing boats discard or lose large amounts of fishing gear, such as fishing lines, buoys, and nets.

A major problem with plastic debris is its persistence. Plastic materials, prized by humans for their inertness and durability, degrade very slowly and most will persist in the environment for many hundreds of years. The oceans have thus been subject to more than half a century of accumulated plastic waste. Plastic debris is now common everywhere in the oceans: floating on the surface and suspended in the water column; accumulating in all five oceanic gyres which, because of their circular motion, tend to trap floating debris; settling on the ocean floor at all depths; and littering all coasts. Over time, plastic debris breaks down into tiny 'microplastic' fragments that are suspended in the oceans and then sink very slowly to the ocean floor.

The amount of plastic debris stranded on shorelines throughout the world is appalling. In the UK there is on average over 7,000 pieces of plastic along a kilometre of shoreline; in the Caribbean about 1,900 to over 11,000 items; in Indonesia more than 29,000. Surveys of the coastal seabed show that there are typically hundreds of items of plastic debris per square kilometre and in some places in Indonesia and the Caribbean there can be thousands of items of plastic per square kilometre.

The amount of plastic debris floating in the oceans was surveyed over the course of twenty-four research voyages that took place between 2007 and 2013 across all five oceanic gyres, along coastal Australia, and in the Bay of Bengal and the Mediterranean Sea. Using the data collected it was estimated that more than 5 trillion pieces of plastic weighing over 250,000 tonnes are floating on the surface of the oceans. Nets towed from the vessels routinely captured 1,000 to 100,000 pieces of plastic per square kilometre of ocean surface and in some tows much more than this. Most of the plastics in the oceans—about 92 per cent—consists of microplastics less than 4.75 mm in size.

Plastic debris is very harmful to many forms of marine life which mistake it for food and ingest it or become entangled in it. Hundreds of different species have been recorded harmed by plastic debris, including seabirds such as penguins, albatrosses, pelicans, and many shorebirds; and marine mammals, including whales, seals, sea lions, sea otters, sharks, manatees, dugongs, and turtles. Entanglement is often caused by discarded fishing nets and ropes, monofilament fishing line, packing strapping bands, and six-pack rings. Seabirds and turtles routinely ingest plastic debris of all sorts, probably mistaking it for items of food. Turtles, for example, appear to mistake plastic bags for jellyfish, one of their prey items. Ingestion of plastic can lead to obstruction of the gut and death, while toxic chemicals can leach out of the ingested material and cause other harmful effects. Microplastics have also been found in the gut and tissues of zooplankton, including copepods. Disturbingly, plastic particles have now been found in seafood eaten by humans, including cod, haddock, mackerel, and shellfish. The effect on humans of eating plastic contaminated seafood is still being debated, although in 2016 the European Food Safety Authority warned of a potential increased risk to food safety and human health.

The scale of the plastic debris crisis is daunting but, thankfully, there is a rapidly growing level of societal awareness and concern about the issue, and initiatives to address the problem are gaining momentum. It is generally agreed that a multi-pronged approach must be pursued involving governments, NGOs, and industry worldwide including: improving waste collection systems, particularly in poorer nations, to reduce loss of plastics from the land to the oceans; greatly increasing the percentage of plastics that are recycled into reused plastic products; reducing the amount of virgin plastic products that are manufactured; stemming the demand for plastics through initiatives such as banning single use plastic bags and reducing plastic food packaging; removing plastic debris from rivers before it reaches the oceans; and removing plastic debris that is already in the marine environment to reduce its ongoing harm to marine organisms and prevent it breaking down into more harmful and impossible to remove microplastics.

The Ocean Cleanup, a non-profit organization, has launched an ambitious plan to significantly reduce the plastic debris concentrated in the North Pacific gyre, which is often referred to as the Great Pacific Garbage Patch. The first working version of their collection system was deployed in the gyre for testing in late 2018. The system consists of a 600-metre-long U-shaped floating tube from which hangs a 3-metre-deep skirt. Wind and waves passively move plastic floating at or just below the surface into the centre of the system where it is concentrated. The skirt is made of smooth, impenetrable material so plankton cannot get trapped in it, and fish can swim underneath the skirt. A modified version of this collection system designed to collect plastic more effectively was deployed in mid-2019 for further testing and is now successfully collecting plastic. Ocean Cleanup now plans to build and deploy a fleet of about sixty full-scale systems, with vessels removing the plastic from the systems every few months and returning it to land for recycling. The organization projects these systems could remove 50 per cent of the plastic in the Great Pacific Garbage Patch in five years. Nonetheless, to adequately address the problem of marine plastics these clean-up systems will have to be a part of a broader suite of initiatives that stem the flow of plastic into the oceans from land.

# Chapter 4
# Polar marine biology

Flourishing marine biological systems are present in the extreme environments of the Arctic and Antarctic polar regions of the planet. Both these regions are characterized by constantly cold sea temperatures, ice-covered oceans, and extreme seasonal fluctuations in light levels. In many other ways, however, these regions are very dissimilar and have evolved strikingly different and unique marine ecosystems.

## Marine biology of the Arctic Ocean

The Arctic Ocean is a comparatively small (about 15.6 million square kilometres) and isolated body of seawater with extensive areas of shallow, continental shelf. It is largely surrounded by land masses with only two outlets, the very narrow Bering Strait to the Pacific Ocean, which is only 70 metres deep; and the broader 400-metre-deep Fram Strait to the Atlantic Ocean. Several large Siberian and Canadian rivers empty into the Arctic Ocean, creating a thin, lower-salinity layer of seawater about 20–50 metres deep that floats on the saltier and denser seawater beneath. Extensive areas of the Arctic Ocean have a soft sedimentary bottom resulting from the discharge of large amounts of sediment from these river systems.

The surface of the Arctic Ocean is at or near the freezing point of seawater (−1.9°C) for much of the year. Thus, much of the Arctic Ocean is permanently covered by a floating cap of sea ice which expands and retreats with the seasons. The cap is largest in April at the end of the Arctic winter and smallest in September at the end of the Arctic summer. The summer melt occurs mostly over the vast continental shelves of the Arctic Ocean, while most of the central Arctic Ocean remains covered by ice throughout the year. This 'multi-year' sea ice has survived complete melting for several years and is 3–4 metres thick. The remainder is thinner first-year ice about 1–2 metres thick.

One might expect this vast volume of frozen seawater to be devoid of life. In reality, it harbours an abundant and diverse community of marine life, unique to polar seas, and one which plays a fundamental role in sustaining the polar food web. Sea ice is habitable because, unlike solid freshwater ice, it is a very porous substance. As sea ice forms, tiny spaces between the ice crystals become filled with a highly saline brine solution resistant to freezing. Through this process a three-dimensional network of brine channels and spaces, ranging from microscopic to several centimetres in size, is created within the sea ice. These channels are physically connected to the seawater beneath the ice.

During the autumn and winter period, viruses, bacteria, archaea, and protists in the seawater become incorporated into the downward growing sea ice. A significant amount of the primary production in the Arctic Ocean, perhaps up to one-third in those areas permanently covered by sea ice, takes place in the ice. In the Arctic spring and summer, enough light penetrates the snow covering the ice, and the ice itself, to sustain primary production by photosynthetic bacteria, diatoms, and dinoflagellates present within the brine channels, and also in a layer coating the bottom of the ice. These organisms soon become so abundant that they colour the ice brown (see Figure 18). The sea ice provides a uniquely stable marine habitat that keeps these photosynthetic

18. Ice core 1 m in diameter showing distinct brown coloration of ice biota.

organisms within the photic zone at all times during the summer, thus maximizing primary productivity.

A well-developed microbial community is present in sea ice that functions in a similar way to the microbial loop of the open-ocean pelagic zone (see Chapter 2). Autotrophic bacteria, along with heterotrophic bacteria feeding on DOM, are eaten by abundant flagellated protists, which are then food for larger ciliated protists.

Large zooplanktonic organisms, such as amphipods and copepods, swarm about on the under surface of the ice and shelter in the brine channels, grazing on the ciliates and larger photosynthetic organisms such as diatoms and dinoflagellates. These larger grazers provide a key link to higher trophic levels in the Arctic food web (see Figure 19). They are an important food source for fish such as Arctic cod that feed along the bottom of the ice. These fish are in turn fed on by squid, seals, and whales. The seals are an important source of food for the roughly 25,000 polar bears that currently inhabit the Arctic region. Polar bears are adept at killing seals as they emerge through breathing holes in the ice, or when they haul themselves up onto the ice edge.

Partway through the summer, as the edge of the sea ice begins to melt, photosynthetic organisms associated with the ice are released into the seawater and seed an under-ice phytoplankton bloom. As the summer progresses, and the ice edge breaks up and retreats, this bloom creates a 20–80-kilometre-wide zone of extremely high productivity at the ice edge. Walrus, seals, narwhal, beluga whales, and bowhead whales are abundant at this ocean–ice boundary, along with seabirds and polar bears. This community follows the ice edge for hundreds of kilometres as it retreats northwards during the Arctic summer.

Unconsumed organic material arising from these intense Arctic blooms sinks to the ocean bottom and supports the Arctic benthic

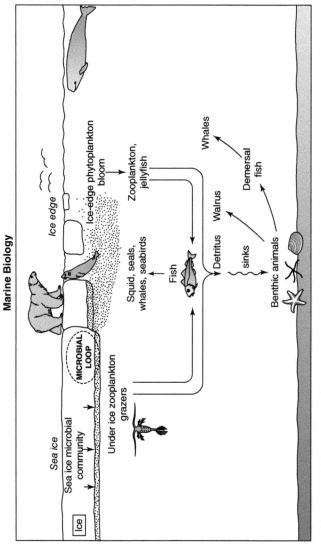

19. Depiction of the Arctic food web.

community. Benthic organisms, such as amphipods, worms, clams, and brittle stars, live on or within the soft-bottom sediments of the Arctic Ocean and are particularly common on the continental shelves where they are fed on by bottom-feeding fish such as eelpouts and sculpins, as well as by grey whales and walruses that forage on the seabed.

As a result of human-induced climate change, the Arctic region is warming, and it is doing so at a rate faster than the rest of the planet. This is having a great impact on the Arctic Ocean ice cap. One stark trend is that the overall thickness of the sea ice is decreasing rapidly—it declined from a mean of 3.64 metres in 1980 to 1.89 metres in 2008. Prior to the late 1970s, sea ice typically extended over close to 15.6 million square kilometres in late winter in the Arctic. However, the maximum winter extent of sea ice in the Arctic Ocean has been decreasing by an average of 3 per cent per decade since 1979 and in recent years at a much faster rate than this; it is now down to about 14.4 million square kilometres. The extent of minimum summer ice cover is decreasing at an even faster rate. Sea ice typically extended over about 7 million square kilometres at summer's end in the Arctic. Over the past decade, though, the minimum sea ice extent has been only 3.5 to 5 million square kilometres. At this rate, the Arctic Ocean will become nearly or completely ice free for several months a year before 2040, and possibly within the next decade.

Clearly, the Arctic Ocean as we have known it is about to disappear with profound effects on its marine biology. For example, overall primary production will increase substantially, since less snow and ice cover means a deeper photic zone; also, much of the sea-ice microbial community will disappear in summer. These changes alone will affect the Arctic food web in profound ways. Furthermore, seals and polar bears, which live in close association with sea ice, will be particularly hard hit, since their feeding and breeding habitat will be greatly reduced.

There are no commercial fisheries in the Arctic Ocean currently because of the difficulty of operating fishing vessels in ice-covered seas. But the Arctic Ocean will become more attractive to commercial fishing fleets as the ice cap shrinks further, particularly since ocean warming is causing some important fish stocks, such as cod and halibut, to move further north into Arctic waters. Fortunately, the countries of the European Union plus nine other nations have recently agreed to a moratorium on fishing across much of the Arctic to at least 2034. This will allow time for marine biologists to better understand the Arctic's changing marine ecosystem and, hopefully, provide guidance on its future sustainable management. The best outcome would be a further agreement to designate the Arctic Ocean as a permanent marine protected area.

## Marine biology of the Southern Ocean

The Arctic and Southern Ocean marine systems can be considered geographic opposites. In contrast to the largely landlocked Arctic Ocean, the Southern Ocean surrounds the Antarctic continental land mass and is in open contact with the Atlantic, Indian, and Pacific oceans. Whereas the Arctic Ocean is strongly influenced by river inputs, the Antarctic continent has no rivers, and so hard-bottomed ocean floor is common in the Southern Ocean, and there is no low-saline surface layer, as in the Arctic Ocean. Also, in contrast to the Arctic Ocean with its shallow, broad continental shelves, the Antarctic continental shelf is very narrow and steep.

The approximate northern boundary of the Southern Ocean is often designated as the 60° S latitude. Since the edge of the Antarctic continent is at roughly 70° S, the Southern Ocean consists of a ring of ocean about 10° of latitude wide—roughly 1,100 kilometres.

The Antarctic continent is covered by a thick ice sheet that flows outwards towards the coasts, and then out into the ocean to form a vast, floating, 100-metre-thick mass of permanent ice called the ice shelf. Seaward of the ice shelf, the ocean freezes seasonally.

The seasonal fluctuations in the Southern Ocean are enormous. At the start of the southern hemisphere winter, the sea begins to freeze, the freezing front moving outwards rapidly at rates of tens to hundreds of kilometres each day. By the end of the southern hemisphere winter, up to 18 million square kilometres of ocean are covered by sea ice. Unlike the Arctic Ocean, almost all this sea ice will melt during the summer, with only about 3 million square kilometres remaining. Since most of the sea ice in the Southern Ocean is just one year old, it is comparatively thinner than in the Arctic Ocean, generally only 1–2 metres thick.

Antarctic waters are extremely nutrient rich, fertilized by a permanent upwelling of seawater that has its origins at the other end of the planet. As described in Chapter 1, cold dense seawater formed in the North Atlantic Ocean—the North Atlantic Deep Water—sinks and flows slowly southward near the bottom of the Atlantic basin, to emerge hundreds of years later off the coast of Antarctica. This continuous upwelling of cold, nutrient-rich seawater, in combination with the long Antarctic summer day length, creates ideal conditions for phytoplankton growth, which drives the productivity of the Southern Ocean.

As in the Arctic, a well-developed microbial community is present in the sea ice. Sea-ice photosynthetic organisms are even more abundant and productive than in the Arctic Ocean because the sea ice is thinner, and there is thus more available light for photosynthesis. Diatoms can be particularly abundant, and as the sea ice breaks up and recedes in the spring, these are released into the open water at the ice edge and seed huge diatom blooms.

This massive pulse of primary production supports the Southern Ocean's most important marine species, the Antarctic krill, *Euphausia superba*. Antarctic krill are shrimp-like, nearly

**20. Antarctic krill.**

transparent, zooplanktonic animals about 4–6 centimetres in length (see Figure 20). They can live for five to ten years and thus are able to survive through successive long, dark Antarctic winters when phytoplankton are absent and other food is scarce. During the winter months they lower their metabolic rate, shrink in body size, and revert to a juvenile state. When food again becomes abundant in the spring, they grow rapidly and redevelop sexual characteristics.

In the Antarctic spring, krill crawl about under the sea ice grazing on the lush lawn of photosynthetic microbes. As the sea ice breaks up, they leave the ice and begin feeding directly on the huge blooms of free-living diatoms using their filter-feeding appendages. With so much food available they grow and reproduce quickly, and start to swarm in large numbers, often at densities in excess of 10,000 individuals per cubic metre—dense enough to colour the seawater a reddish-brown.

Krill swarms are patchy and vary greatly in size, from some that can be quite small—just patches of a few square metres when looking down on the ocean surface—to ones that are hundreds of square kilometres in size and contain millions of tonnes of krill. Because the Southern Ocean covers a large area, krill numbers are enormous, estimated at about 600 billion animals on average, or about 500 million tonnes of krill. This makes Antarctic krill one of the most abundant animal species on the planet in terms of total numbers and biomass. For comparison, the human population is about 7.3 billion currently, with a total biomass roughly that of Antarctic krill. It is not surprising, then, that krill play a pivotal role in the Southern Ocean marine system. They are a staple food source for almost all the Southern Ocean's marine animals including fish, squid, leopard seals, fur seals, crabeater seals, penguins, seabirds, and baleen whales. They are thus a key link in a short and efficient two- or three-step food chain (see Figure 21).

Blue, right, and fin whales are abundant in Southern Ocean waters during the summer. These whales have sieve-like baleen plates in their mouths which they use to capture large quantities of krill. They take in large mouthfuls of krill-laden seawater and then use their tongue to force the seawater out through the baleen sieve, which retains the krill.

All Southern Ocean whale species are migratory, feeding in the Southern Ocean during the southern hemisphere summer, and then swimming long distances to warmer northern waters to breed during the winter months. In the southern hemisphere spring they head back south, following the retreating ice edge.

Crabeater seals, despite their name, also depend on krill as a food source. These seals, which live on ice floes, have specialized teeth that are adapted for straining krill directly from the ocean. Much like baleen whales, they take a mouthful of seawater and expel it through their teeth, which retain the krill. Not surprisingly, given the huge size of their food stock, crabeater seals are very abundant although accurate estimates of their numbers remain

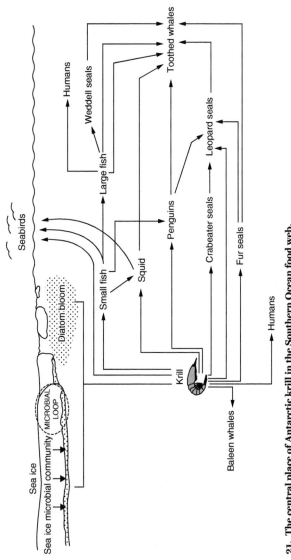

21. **The central place of Antarctic krill in the Southern Ocean food web.**

elusive. It is currently thought that there are at least 7 million individuals, which would make them the most abundant seal species on the planet.

Antarctic penguins also rely on krill. The most abundant species of penguin in Antarctica is the Adelie penguin. There are about 2.5 million breeding pairs, which feed on krill and small fish. They are capable of diving to depths of hundreds of metres in pursuit of food. The young penguins are especially dependent on krill, and if krill numbers are low in any particular year, juvenile mortality is high.

Large marine predators are common in the Southern Ocean. Leopard seals are voracious carnivores well equipped to feed on large prey, such as penguins, crabeater seals, and fur seals. Yet they too can feed on the ubiquitous Antarctic krill and, like the crabeater seal, have some of their teeth modified to act as krill strainers. Killer whales, or orcas, are another common Antarctic predator, feeding on fish, penguins, seals, and other whales, but even orcas will feed on krill.

Many different species of squid are common in the Southern Ocean and are a very important food source for large marine animals, including toothed whales, such as sperm whales, and seabirds. One of the largest invertebrates on the planet, the colossal squid, inhabits the deep waters of the Southern Ocean. Until recently, a complete specimen of this animal had never been seen, and it was known only from bits found in the stomachs of sperm whales hunted down by whalers. But in 2007 a live colossal squid was brought to the surface from a depth of around 2,000 metres in the trawl net of a deep-ocean fishing vessel from New Zealand. This animal was 10 metres in length and weighed close to 500 kilograms. Stomach analyses show that colossal squids are the major prey of sperm whales in the Southern Ocean. Many sperm whales have scars on their body which appear to be caused by the sharp hooks present at the tips of the colossal squid's two long feeding tentacles, evidence of mighty predator–prey struggles in the cold abyssal depths of the Southern Ocean.

An unusual group of fish, the icefish, are common in Antarctica. These fish, which live constantly in seawater which is on the verge of freezing, have very little of the oxygen transporting red pigment, haemoglobin, in their blood streams; oxygen is simply transported in solution in their blood plasma. These fish can obtain enough oxygen in this fashion because their body fluids are very cold and the amount of oxygen in solution increases with decreasing temperature. They also have other adaptations that prevent their bodies from freezing. For example, their body fluids contain complex proteins and sugars that provide a kind of antifreeze protection by lowering their freezing point.

Seabirds, including albatrosses, petrels, and fulmars, range widely in the Southern Ocean, feeding on krill, squid, and fish. The Wandering Albatross (*Diomedea exulans*) is perhaps the quintessential Southern Ocean seabird. It is the largest of the albatrosses, with a wingspan of 3 metres or more, and spends most of its life on the wing, gliding on the Southern Ocean wind systems. Albatrosses will range over thousands of square kilometres of ocean foraging for food and can travel up to 1,000 kilometres in a day. Albatrosses return to land to breed, mainly on subantarctic islands.

Despite the extreme cold, the benthic communities of the Southern Ocean can be extraordinarily rich. In shallow waters less than about 15 metres, the ocean floor is routinely scoured by grounded ice. Here attached marine animals are absent, and the bottom is occupied during the ice-free period by mobile animals such as sea stars, sea urchins, and large nemertean worms that invade the area when possible to feed on diatoms growing on the ocean floor and to scavenge on dead and dying animals.

In deeper water, attached benthic animals such as sea anemones, corals, and sponges proliferate in great numbers. In some areas, the densities of these ocean floor animals are among the greatest recorded in any marine environment on the planet.

Southern Ocean marine mammals have been ruthlessly exploited in the past. Hunting of Antarctic fur seals began in the late 1700s, and by 1830 most fur seal colonies had been exterminated or reduced to a size where it was uneconomical to continue to hunt them. They were declared a protected species in 1964 and they now number more than five million individuals, which may be a larger population than when exploitation first began.

Southern Ocean whaling commenced in the early 1900s, initially targeting humpback whales, and then spreading rapidly to other species, including southern right, blue, fin, sei, and sperm whales. In the years before the Second World War, tens of thousands of whales were taken annually and in the period between 1956 and 1965, 631,518 whales were recorded killed. The industry collapsed in the 1960s when it became uneconomical to hunt the remnant populations that remained. By that time, the southern right and humpback whale populations had been reduced to about 3 per cent of their original size, blue whales to about 5 per cent, and fin and sei whales to about 20 per cent. A moratorium on commercial whaling came into force in 1987 and, apart from some whaling that Japan considers has a scientific value, no whales are now exploited.

Following the demise of seal and whale populations, the exploitation of Southern Ocean marine species moved down the food web to target smaller animals at lower trophic levels. Commercial fishing began in the late 1960s, first focused on species such as the mackerel icefish. In the 1980s, fishing vessels began to exploit the Patagonian toothfish using longlines set at depths of over 1,000 metres. This fish can grow to over 2.3 metres and weigh more than 130 kilograms. It is marketed as 'Chilean sea bass' and has become a very popular fish commanding a premium price. Catches are currently recorded at about 12,000 tonnes per year. In the late 1990s, this longline fishery expanded to a related species, the Antarctic toothfish, with landings currently about 4,000 tonnes per year. Toothfish are food for sperm whales, killer whales, Weddell seals, and large squid, so their removal will potentially impact on these dependent species. Demand for toothfish outstrips supply, which has led to illegal fishing by vessels which have not been assigned a quota for these species. The value of the illegal catch is in the hundreds of millions of US dollars annually, although in recent years the amount of illegal catch has been reducing due to management measures implemented by the Commission for the Conservation of Antarctic Marine Living Resources (CCAMLR).

Even the krill in the Southern Ocean are subject to human exploitation. The krill are caught by trawling down to about 200 metres. Antarctic krill fishing began in the 1970s and by the early 1980s about half a million tonnes were being harvested annually. Catches then declined from the early 1990s to around 100,000 tonnes per year as most nations abandoned the fishery because of the high cost of operating in the Southern Ocean. But the demand for krill is increasing again and since 2010 catches have been between about 200,000 and 300,000 tonnes per year. Much of the krill is processed into fish meal, which is a component of the artificial feeds used on fish farms. More recently, krill is in demand as a source of health food supplements, such as omega-3 oils.

Krill numbers have declined significantly in some parts of the Southern Ocean since the 1970s, perhaps by as much as 80 per cent. This appears to be the result of rising air temperatures in the Antarctic causing strong declining trends in sea-ice extent in some regions. Krill rely on the sea ice for shelter and food during the winter months, and it has been observed that in years when the extent of sea ice is low, krill are less abundant in subsequent years. It also appears that as krill stocks decrease, salp numbers increase. Salps are planktonic, gelatinous animals that can live in warmer, less productive waters than krill.

The long-standing exploitation of Antarctic marine resources and the growing impacts of climate change belie the commonly held notion that the Southern Ocean possesses one of the planet's last pristine marine systems. Human impacts have undoubtedly pushed the Southern Ocean marine system out of its natural equilibrium, although it has been difficult to document the changes because of the region's remoteness, the lack of a historical baseline of what was 'natural' in the Southern Ocean, and the complexity of species interactions.

An important first step in protecting and restoring the Southern Ocean marine system took place in 2017 with the establishment of the Ross Sea Region Marine Protected Area. The Ross Sea, a deep bay in the Southern Ocean, now contains the largest marine protected area in the world at 2.1 million square kilometres. No fishing is allowed in 1.1 million square kilometres of this reserve. Unfortunately, the agreement expires in 35 years so this is not yet a permanent marine protected area. In 2018, a bid by CCAMLR members to create an even larger marine protected area in the Weddell Sea region of the Southern Ocean was blocked by Russia, China, and Norway, and in 2019 a proposal to create a marine protected area off the east coast of Antarctica was also rejected. There is obviously still a long way to go before enough of the Southern Ocean is properly protected from future exploitation to allow the recovery of this important marine environment.

# Chapter 5
# Marine life in the tropics

The tropical marine environment encompasses those parts of the Global Ocean where the surface waters are consistently warm throughout the year, rarely falling below 20°C. Such regions occur within an oceanic belt straddling the equator from roughly the Tropic of Cancer in the northern hemisphere to the Tropic of Capricorn in the southern hemisphere (23° N latitude to 23° S latitude).

## Coral reefs

Coral reefs embody the archetypal image of a tropical marine environment and are globally significant natural systems in terms of their beauty, biological diversity, productivity, and economic significance (see Figure 22). These 'rainforests of the ocean' are very complex systems that are home to an incredible diversity of marine organisms—one-quarter to one-third of all marine species—with the number of different species on coral reefs globally totalling in the millions.

Coral reefs provide food for hundreds of millions of people, with reef fish species comprising about one-quarter of the total fish catch in less developed countries. They serve as natural protective

**22. Aerial view of coral reefs of Heron Island, Great Barrier Reef, Australia.**

barriers, sheltering coastal communities from the waves generated by hurricanes, typhoons, and cyclones. They are also the basis of employment through tourism for millions of people in the many regions with reefs in their coastal waters. Apart from these ecosystem services, valued in many billions of dollars, coral reefs have tremendous intrinsic value that is impossible to quantify as anyone who has snorkelled or dived on a healthy reef can attest—without coral reefs our planet and human society would be infinitely poorer.

## Physical requirements of corals

Notwithstanding their importance, coral reefs occupy a very small proportion of the planet's surface—a little over 284,000 square kilometres—less than 1.2 per cent of the continental shelf area of the oceans. Coral reefs are thus a rare ecosystem at the global scale. This is because the physical requirements of the main reef-building animals—the corals—are very specific.

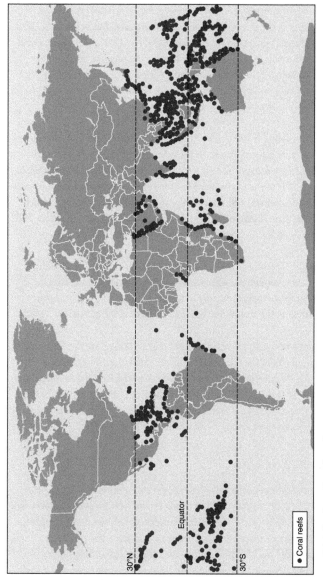

23. Distribution of coral reefs.

Reef-building corals thrive best at ocean temperatures above about 23°C and few exist where temperatures fall below 18°C for significant periods of time. Thus, coral reefs are absent at tropical latitudes where upwelling of cold seawater occurs, such as the west coasts of South America and Africa. Corals also require high light levels to thrive, so they are generally restricted to areas of clear water less than about 50 metres deep.

Reef-building corals generally cannot tolerate freshening of seawater below a salinity of about 30 for extended periods and so do not occur in areas exposed to intermittent influxes of freshwater, such as near the mouths of rivers, or in areas where there are high amounts of freshwater run-off. Therefore, coral reefs are absent along much of the tropical Atlantic coast of South America, which is exposed to freshwater discharge from the Amazon and Orinoco rivers.

Finally, reef-building corals flourish best in areas with moderate to high wave action, which keeps the seawater well aerated, brings in a constant supply of food for the corals, and removes light-blocking sediment from the surface of the corals.

Spectacular and productive coral reef systems have developed in those parts of the Global Ocean where this combination of physical conditions converges, such as in the Caribbean Sea; the many islands of Indonesia, the Philippines, the South Pacific, and the tropical Indian Ocean; in the Red Sea; and off the north-east and north-west coasts of Australia (see Figure 23).

## Biology of corals

Reef-building corals, also known as scleractinian or stony corals, are colonial animals related to sea anemones. Each colony consists of thousands of individual animals called polyps (see Figure 24). Colonies grow through asexual reproduction—the polyps repeatedly bud off new polyps, creating an expanding layer of genetically identical polyps which share a common stomach cavity.

As the colony grows, the polyps extract calcium from the surrounding seawater to secrete a sizeable calcium carbonate skeleton which is external to the polyps themselves. Depending on the species, polyps are positioned within individual cups in the skeleton, or in rows within long grooves in the skeleton. The polyps can retract into the skeleton for protection.

One of the remarkable attributes of reef-building corals is that, although they are animals, functionally they perform in many ways like plants, which explains why they only flourish in well-lit environments. This is because all reef-building corals have entered an intimate relationship with photosynthetic microbes. The tissues lining the inside of the tentacles and stomach cavity of the polyps are packed with photosynthetic cells called zooxanthellae (see Figure 24). These are modified photosynthetic dinoflagellates, a group of phytoplankton generally found free living in the oceans

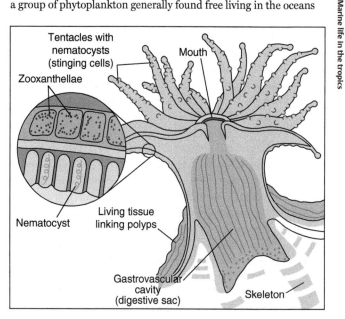

**24. Anatomy of a coral polyp.**

(see Chapter 2). One square centimetre of coral tissue can contain several million zooxanthellae cells.

Reef-building corals 'cultivate' the zooxanthellae for food. They do not directly consume the zooxanthellae—instead the corals chemically control the density of the zooxanthellae in their tissues and stimulate the zooxanthellae to secrete some of the organic compounds that they synthesize through photosynthesis directly into their gut tissues. Depending on the species, corals receive somewhere between 50 per cent to 95 per cent of their food from their zooxanthellae.

The corals acquire their zooxanthellae in several ways. When the polyps bud asexually, each new polyp retains some zooxanthellae. Coral polyps can also reproduce sexually in which case the polyp often incorporates some of its zooxanthellae into each egg it produces. However, many corals do not inherit zooxanthellae and must acquire them from the surrounding environment as they grow. In this case, it appears that coral larvae secrete a chemical into the seawater that attracts the preferred strains of dinoflagellates, which are ingested and incorporated into the coral's own cells. The coral then surrounds each dinoflagellate cell with a special membrane and begins to control its metabolism.

Both the coral and the zooxanthellae benefit from their symbiotic relationship, although the coral is the controlling party. The beauty of the relationship lies in the way it allows very efficient nutrient recycling between both parties. The photosynthetic zooxanthellae, protectively sheltered within the coral's tissues, acquire a constant supply of waste metabolic products essential for photosynthesis—carbon dioxide, nitrogen, and phosphorus—directly from their coral host. In the presence of light they turn these nutrients into organic compounds, some of which the coral 'steals' back as food. The corals also make use of the oxygen generated by the zooxanthellae as a by-product of photosynthesis.

Although the zooxanthellae provide a large proportion of the coral's energy needs, most reef-building corals supplement their diet by capturing food from the external environment. Corals generally feed at night by extending their polyps above the skeleton, which is why coral colonies appear 'furry' at night. The mouth of each polyp is surrounded by a ring of tentacles equipped with special 'stinging' cells, called nematocysts (see Figure 24), that eject both poisonous and sticky threads that subdue small animals, mainly zooplankton, that the coral colonies feed on. Corals also secrete threads of sticky mucus that collect small organic particles, which are then drawn into the polyps' mouths.

## Types of coral reefs

Corals grow slowly, in the order of a few centimetres per year, but over long periods of time form massive durable structures—the largest made by living organisms. There are three main types of coral reef structures—atolls, fringing reefs, and barrier reefs (see Figure 25).

Atolls are common in the tropical Indian and Pacific Oceans and are associated with oceanic islands (see Figure 25(a)). The creation of an atoll is initiated when reef-building corals colonize the sides of a newly created volcanic island to form a 'fringing' reef (see Figure 25(b)). Such newly formed islands often slowly sink because of the massive weight they exert on the underlying ocean floor. As they sink, the corals surrounding the island continue to grow upwards on a bed of calcium carbonate that they secrete—a platform of limestone that gets deeper and deeper over time. These reefs eventually become separated from the island by a seawater-filled lagoon and are known as 'barrier' reefs (see Figure 25(c)). As the island itself eventually disappears beneath the surface of the ocean, the coral reef continues to grow upwards from its island base towards the ocean surface, creating a ring- or semi-ring-shaped structure around the lagoon—the atoll reef (see Figure 25(d)).

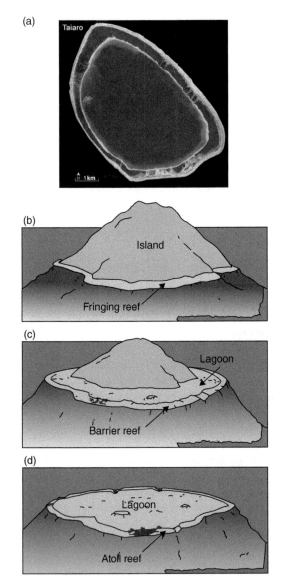

25. Stages in the formation of an atoll reef. The photograph is an aerial view of Taiaro, a small atoll in the west of the Tuamotu group in French Polynesia.

The living crown of an atoll can rest on an enormously thick layer of coral-created limestone. Bore holes drilled at Eniwetok Atoll in the Marshall Islands penetrated close to 1,400 metres of limestone before hitting volcanic rock marking the top of the volcanic island on which the atoll originated. It would have taken about 60 million years of reef growth to create a limestone cap of this thickness.

Fringing and barrier reefs can also develop alongside continental land masses. Fringing reefs are separated from the coast by a narrow channel, whereas barrier reefs occur at a larger distance off the coast. Such barrier reefs develop when the coastline on which they are growing subsides or is flooded by rising sea levels. Barrier reefs can be very large structures, the largest being the Great Barrier Reef which stretches for about 2,600 kilometres off the north-east coast of Australia.

## Productivity of coral reefs

The corals are the backbone of the reef ecosystem, creating a complex, three-dimensional habitat that supports a truly remarkable diversity and abundance of marine life. Marine invertebrates are prolific. Some, such as sponges, sea fans, and soft corals, live attached to the reef. Others are more mobile, such as sea urchins, sea cucumbers, sea stars, crabs, shrimps, and sea slugs. Colourful fish are abundant and conspicuous, and on a healthy reef large predatory fish such as groupers, barracuda, and sharks are common.

Healthy coral reefs are very productive marine systems. This is in stark contrast to the nutrient-poor and unproductive tropical waters surrounding the reefs. Coral reefs can be 50 or even 100 times more productive than the surrounding oceanic environment, and for this reason are often referred to as oases in a tropical marine desert.

It is at first difficult to comprehend how coral reefs can be so productive since there are no obvious primary producers on a reef. However, the zooxanthellae hidden within the tissues of the corals themselves occupy up to 10 per cent of the biomass of living corals and thus represent a substantial mass of microbial primary producers. Other kinds of primary producers are also widespread on a coral reef, including microscopic algae and cyanobacteria that bore into coral skeletons, red and green coralline algae that form a ubiquitous encrusting layer on exposed hard surfaces, and various forms of delicate calcareous algae that form turf-like layers on parts of the reef. All this adds up to a large, although somewhat inconspicuous, mass of photosynthetic organisms on the reef.

The seawater flowing over a coral reef is well lit but very nutrient poor. So where do the nutrients come from to support these highly productive tropical marine oases? As it turns out, coral reefs are superb nutrient sinks, able to scavenge the available nitrogen and phosphorus from their nutrient-poor surroundings, and then retain and use these nutrients very efficiently. Reef algae absorb some of the scarce nutrients directly from the seawater flowing over the reef. In addition, the corals obtain some nutrients from the zooplankton and dead organic particles that they filter from the seawater as a supplementary food source. Nitrate is also created by nitrogen-fixing cyanobacteria living in association with the corals and other reef organisms, such as sponges, or free living in the seawater. Once acquired, these precious nutrients are recycled very efficiently back and forth between the primary producers and animals on the reef, with minimal loss to the surroundings. The ultimate example of this is the tight recycling of nutrients between zooxanthellae and the tissues of their coral hosts.

Exposed fleshy macroalgae, or seaweeds, occur sparsely on a healthy coral reef because they are readily consumed by herbivorous fish, such as damselfish and surgeonfish, which are normally very abundant on a reef. Sea urchins, such as the menacing-looking, long-spined, black sea urchin *Diadema*, are also efficient grazers. Together, these herbivores play a very important role in maintaining a healthy coral reef system by preventing the otherwise fast-growing macroalgae from overgrowing and killing the corals by blocking out their light.

Corals are not completely immune to predation despite their protective external skeletons. Several types of fish, known as corallivores, are well adapted to feeding on corals. Some, such as butterfly fish, pluck entire coral polyps from a colony. The forceps-like mouths of butterfly fish, which are armed with numerous small teeth, are well suited for this. Others, such as parrotfish, bite or rasp off and ingest pieces of coral, together with the skeleton, and digest the soft tissues of the coral as well as the algae and bacteria in the coral skeleton. The beak-like mouths of parrotfish, consisting of fifteen rows of extremely hard teeth, are well adapted for this sort of feeding behaviour. When snorkelling or diving on a reef, one can hear the loud noise made by a school of parrotfish crunching on coral. The remains of the material digested by parrotfish are excreted as sand. One large parrotfish produces about 450 kilograms of sand in a year which accumulates in pockets on the reef and helps form the sandy beaches associated with some reef systems.

## Sexual reproduction in corals

Reef-building corals can disperse from the parent colony and colonize new habitats through sexual reproduction. Most species of coral are hermaphroditic—able to produce both eggs and sperm on the same colony. In some species, separate male and female colonies are the norm. Most coral species employ what is termed broadcast spawning—they eject huge numbers of sperm and eggs into the ocean where fertilization takes place. The fertilized eggs develop into tiny ciliated larval forms, called planulae, which are transported by currents for days or weeks, depending on the species. When the planulae detect favourable conditions they swim to the bottom, where they attach themselves and start a new colony.

On coral reefs throughout the world, the coral colonies over a large portion of a reef often spawn simultaneously in a spectacular mass spawning event in which the surrounding seawater becomes saturated with coral gametes, which form distinct slicks on the ocean surface. On parts of the Great Barrier Reef, for instance, millions of colonies consisting of many different species of coral spawn together on a single night, or over the course of a few nights, during the southern hemisphere spring or early summer after a full moon. Mass spawning in corals has most likely evolved to create very dense concentrations of sperm and eggs in the seawater to ensure high rates of successful fertilization. The coral eggs can distinguish among the different kinds of sperm present in the seawater during a mass spawning, rejecting sperm of a different species, and so limit the chances of hybridization between species.

The breeding season of coral colonies is controlled by factors such as seasonal changes in ocean temperature or day length, which serve to bring the corals into breeding condition at the same time. The actual spawning of the ripe colonies is triggered by different factors, including lunar periodicity, a decrease in light levels at sunset, and chemical cues released into the water by other colonies of the same species. So-called 'primitive' corals can thus detect light levels, distinguish the phases of the moon, and communicate with each other chemically.

## Physical and biological disturbances on coral reefs

Though massive in structure, coral reefs are subject to large-scale physical disturbances. The large waves created by hurricanes, typhoons, and cyclones passing in the vicinity of a reef routinely break up and overturn extensive areas of living coral. The rate of recovery from damage of this kind is generally in the order of decades or longer. Ocean warming caused by climate change is creating the conditions for more frequent and powerful storms that are causing greater, more regular and widespread damage to coral reefs and limiting their ability to fully recover between successive storm events. Large tsunami waves generated by major earthquakes, such as the 26 December 2004 earthquake near Indonesia, also cause large-scale damage. Furthermore, coastal reefs are periodically inundated by freshwater run-off following major flood events, which can kill large numbers of corals.

Coral reefs are also susceptible to large-scale biological disturbance. Population explosions of the 'crown-of-thorns' sea star of the genus *Acanthaster* routinely result in the destruction of coral reefs in the Pacific and Indian Oceans and in the Red Sea. The crown-of-thorns is a very large sea star, reaching a diameter of half a metre, and is a specialist feeder on coral polyps. It normally occurs at very low densities on a reef, less than one animal per hectare. Each sea star on its own can strip the tissue from a square metre or so of living coral a month, which is generally not enough to harm the reef. But when densities exceed about thirty sea stars per hectare of reef, the sea stars begin to consume corals at a rate faster than they can grow, and the results can be disastrous for the reef. During a crown-of-thorns outbreak, which can last for many years, with the sea stars roving from one reef to the next and reaching densities of over 1,000 individuals per hectare, large areas of coral are denuded, leaving behind bare skeletons (see Figure 26). Recovery following such an outbreak is a slow process taking anywhere from about 5 to over 100 years depending on the extent and severity of coral loss, and reefs can be re-infested with crown-of-thorns before they are completely recovered.

Crown-of-thorns outbreaks were first documented in the 1960s in Japan and Australia and have been observed regularly since then in many different places. Such outbreaks probably occurred naturally in the past, but the frequency and size of outbreaks are

26. Crown-of-thorns sea star feeding on coral colonies. The corals to the right of the picture have been stripped of their polyps by the sea star, leaving behind bare skeleton.

now increasing, which suggests a human influence is involved. The crown-of-thorns' array of venomous spines makes it an unpalatable meal for most potential predators, although several species of fish can eat it, as well as a large marine snail, the giant triton. It has been proposed that overfishing of some of its fish predators, together with overzealous collection of the giant triton, whose shell is prized by souvenir hunters and shell collectors, has allowed crown-of-thorns populations to explode in some areas. It has also been suggested that such population explosions are linked to run-off from land following abnormally heavy rainfall, which flushes excessive nutrients from agricultural land into coastal waters. The nutrients stimulate blooms of phytoplankton which create a plentiful source of food for the young planktonic larval stage of the crown-of-thorns. This may result in unusually high rates of survival of the young stages leading to a pulse of high juvenile recruitment and a population explosion some years later.

Though the causes of crown-of-thorns outbreaks are yet to be fully understood, they are causing considerable damage to coral reefs in many areas. The Great Barrier Reef has been subject to waves of crown-of-thorns outbreaks, the first in 1962, then 1979–91, 1993–2005, and the latest which began in 2010. Coral cover on the Great Barrier Reef has declined by about 50 per cent since 1985 and almost half of this decline has been attributed to crown-of-thorns outbreaks.

The Australian government has attempted to control crown-of-thorns outbreaks on selected reefs using teams of divers who inject the sea stars with a toxin, but this approach is very inefficient and a better way of culling large numbers of sea stars is needed. A new approach involving the discovery of a unique chemical cue that crown-of-thorns sea stars release into the seawater and that prompts them to cluster together prior to spawning may offer some hope. It is possible that this chemical could be developed into a 'bait' to attract large numbers of the sea stars to one spot and make it easier to cull them.

Corals are also affected by a range of diseases which can cause discoloration of the coral tissues, tumours, and tissue death. Little is known about the causes and effects of these diseases, although they are most likely associated with infections by various viruses, bacteria, and fungi. A very serious coral disease known as 'white band disease' has been killing two species of branching corals in the Caribbean—staghorn and elkhorn corals—since outbreaks were noted in 1979. These corals once formed vast, impenetrable, and beautiful thickets in shallow waters throughout the Caribbean, but over the last thirty-five years white band disease has killed up to 95 per cent of them, and both species are now listed on the US Endangered Species Act. Intriguingly, recent research has revealed that a small proportion of the remaining staghorn corals are now resistant to the disease, which opens the possibility of outplanting these disease resistant strains to re-establish populations in selected areas in the Caribbean.

## Local- and regional-scale human impacts on coral reefs

Although subject to a range of natural physical and biological disturbances, reef-building corals have persisted for many millions of years. Now, unfortunately, coral reefs are under serious threat from a wide range of human disturbances at local and regional levels.

Overfishing constitutes a significant threat to coral reefs. About an eighth of the world's population—roughly 875 million people—live within 100 kilometres of a coral reef. Most of these people live in less developed countries and island nations and depend greatly on fish obtained from local coral reefs as a food source. It is not surprising, then, that unsustainable fishing is a rampant problem on most coral reef systems around the planet.

The larger, high-value, predatory fish, such as groupers, snappers, trevally, and humphead wrasse, are the first to be targeted by fishers on a healthy coral reef. These rapidly become depleted and fishers then start to fish down the food chain out of necessity and target mainly herbivorous fish. The result is a coral reef inhabited by small fish difficult to catch and of little food value, and which is virtually devoid of the large predatory species such as sharks and groupers that would have once been common, as well as the larger herbivorous fish that feed on macroalgae.

Coral reefs in this state become less resilient and much more vulnerable to other perturbations in the system. The coral reefs of the Caribbean region provide a classic example of the destabilizing effects of overfishing. Most of the Caribbean island nations are highly populated and many reefs were overfished by the 19th century. Now, at least 60 per cent of the region's coral reefs are severely overfished and large predatory and herbivorous fish are very rare.

At first, the removal of herbivorous fish from the system was compensated by an increase in the numbers of *Diadema antillarum* sea urchins which did not have to compete with herbivorous fish for macroalgal food. These urchins continued to graze back and control the amount of macroalgae on the reef. Then, beginning in 1983, a sea urchin disease spread rapidly through the Caribbean and killed almost all *Diadema* throughout the region within a couple of years. Now that virtually all coral reef herbivores were absent, macroalgae flourished and quickly overgrew the corals. As a result, over the course of a decade, the coral reefs throughout most of the Caribbean were transformed from coral-dominated structures to drab seaweed-dominated systems lacking the colour, diversity, and complexity of a healthy coral reef. Once fleshy macroalgae become established on a coral reef, the regrowth of corals is severely disrupted. Thus, unfortunately, this massive collapse of the Caribbean coral reef ecosystem will persist indefinitely in the face of continued overfishing, compounded by other stresses. Coral reefs around the world are subject to similar overfishing pressures and transformations similar to those that took place in the Caribbean are now widespread.

Corals are particularly susceptible to any degradation of water quality arising from coastal development and land-use change. Sediment run-off from agricultural land, deforested areas, and from earthworks during coastal development decreases water clarity and coats corals with sediment, reducing the amount of light the corals receive, and smothering the polyps. Even very small increases in nutrient concentrations can stress corals by stimulating elevated levels of phytoplankton, which reduce water clarity and light penetration. Increased nutrients also encourage the growth of coral-smothering macroalgae. Untreated sewage discharge is thus an obvious threat; so too are nutrients in agricultural run-off.

Run-off from agricultural land is affecting the resilience of parts of the Great Barrier Reef. Sediment, nutrients, and herbicides drain into the Great Barrier Reef system from a huge catchment area of about 424,000 square kilometres. Cattle grazing takes place throughout much of this catchment and sugar cane is grown on other parts of the catchment, particularly adjacent to waterways on fertile coastal floodplains. It is estimated that sediment run-off is now 5 to 9 times greater and the impact of nutrient-enriched water 10 to 20 times greater than when the catchment was undeveloped prior to 1850.

It is difficult to assess the overall effects of this run-off on the Great Barrier Reef because there is no information on what the reef was like prior to the development of the catchment. It is very likely, though, that reefs within about 10 kilometres of the coast are now at risk from nutrient enrichment and that reefs further offshore are being affected in ways that will soon reduce their resilience to other stressors. In response, efforts are under way to implement new land management practices that will reduce the pollutants in run-off, such as more efficient use of fertilizers, limiting the use of herbicides, and re-establishing riparian vegetation along the edges of rivers and streams to help filter out sediments and nutrients.

Coral reefs globally have only been seriously studied since the 1970s, which in most cases was well after human impacts had commenced. This makes it difficult to define what constitutes a 'natural' and healthy coral reef system, as would have existed prior to extensive human effects. Coral reef biologists have attempted to obtain a clearer picture of what an un-impacted coral reef system is like, and in so doing reset our biased 'baseline', by studying reefs on uninhabited atolls in the remote Line Islands, which are situated in the central Pacific Ocean 1,600 kilometres south of Hawaii. They then compared what they found on the uninhabited atolls with increasingly populated atolls in the same chain of islands which have been subjected to various levels of fishing pressure and pollution.

They found that the coral reefs on the uninhabited atolls are dominated by large numbers of top predators comprising large fish such as sharks, jacks, red snappers, and groupers; live corals cover nearly 100 per cent of the bottom, and macroalgae are virtually absent. On these reefs an extraordinary 85 per cent of the fish biomass consists of large predators, about three-quarters of which are sharks. This highly 'top heavy' biomass of predatory fish is sustained by a rapid turnover of quickly reproducing and growing prey fish such as butterfly fish, parrotfish, and damselfish. Coral reefs on the populated atolls are quite different. Here predatory fish are rare, and the reefs are dominated by large numbers of small, aquarium-sized herbivorous fish—the 'bottom heavy' pattern of fish biomass that we have come to think of as characteristic of coral reefs. These reefs have much lower levels of coral cover and much larger amounts of macroalgae. Thus, these few remnant coral reef systems on remote uninhabited atolls provide us with a glimpse of what most coral reefs looked like hundreds of years ago prior to widespread human influence.

## Global-scale human impacts on coral reefs

The human threats to coral reefs discussed thus far are local or regional in scale. The ultimate peril to coral reefs is global in scale—ocean warming and acidification resulting from the human-induced climate crisis.

Corals are very sensitive to sea temperature, and small increases above normal summer maximum temperatures result in stress. Temperature-stressed corals undergo 'bleaching', in which they expel the zooxanthellae from their tissues. Without the zooxanthellae, the coral tissues become transparent and reveal the white limestone skeleton beneath. If temperature stress is of moderate and short duration, corals can reacquire their zooxanthellae and survive, although they may be more susceptible to other stresses, such as disease. Highly temperature-stressed corals are not able to reacquire zooxanthellae quickly enough and without them they die.

Occasional, small-scale episodes of coral bleaching are a natural phenomenon on coral reefs. Starting in 1980, though, global coral bleaching events have been occurring with rapidly increasing frequency and intensity as a result of ocean warming caused by increases in the concentration of carbon dioxide in the Earth's atmosphere (see Figure 27). Since the 1980 event, global scale bleaching has occurred in 1998, 2010, and 2015–17. The last event—the first yearly 'back to back' event, and the longest, most extensive, and most damaging global bleaching event to date—affected every major coral reef region on the planet. Its impact on the Great Barrier Reef was catastrophic, with 49 per cent of all the corals on a 1,600 km stretch of the reef dying. In the past, coral reefs have had time to regain some coral cover prior to another bleaching event. However, modelling of future ocean surface temperatures, combined with knowledge of coral physiology, shows that by about 2050 severe global coral bleaching will be occurring annually, with reefs in some regions experiencing this by the mid-2030s. We are thus now, unfortunately, entering a period in which the frequency of bleaching events is outpacing the capacity of coral reefs to recover.

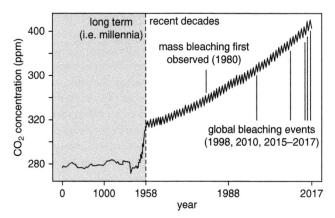

27. **Changes in the concentration of carbon dioxide ($CO_2$) in the Earth's atmosphere over the previous two millennia in relation to the timing of global coral bleaching events. The grey panel shows the long-term trend in $CO_2$ concentration while the white panel shows the trend over recent decades.**

The increasing concentrations of carbon dioxide in the atmosphere are not only causing ocean warming but, as discussed in Chapter 1, are also making seawater more acidic. Ocean acidification makes it more difficult for corals to manufacture their calcium carbonate skeletons. The corals thus must expend more energy to produce their skeletons, which slows their growth and causes stress, making them more susceptible to other stresses such as temperature and disease. Continued acidification will eventually cause corals to stop growing altogether and, at some point, cause their skeletons to slowly dissolve. At current rates of acidification, this will occur around 2080 at which time the skeletons of most corals will start to dissolve faster than they can be built.

## The future of coral reefs

In the face of such a variety of human-related regional and planet-wide impacts, the health of coral reef ecosystems is deteriorating rapidly, and the future is very bleak. Coral dominated ecosystems will disappear entirely from the planet in the next fifty years if the greenhouse gas emission reduction goals of the 2015 Paris Agreement—to keep the increase in average global temperature to well below 2°C above pre-industrial levels—are not met. At current rates of global emissions, the world will soon pass the point at which this goal can be achieved. Even if it is achieved, 70–90 per cent of the planet's corals will disappear by about 2050. We are thus, sadly, at the point where soon coral reefs will become legendary ecosystems with only movies and computer-generated simulations to remind future generations of what has been lost.

Fortunately, passionate coral biologists, philanthropists, and NGOs are now working together to develop long-term contingency plans to conserve at least some coral reef ecosystems in a relatively natural state and, perhaps, to re-establish them at some future time. Several complementary strategies are emerging which, taken together, provide a glimmer of hope for the future of coral reefs.

One important initiative is to establish a global portfolio of reef ecosystems that, on account of their geographic location, are most likely to be least affected by the impacts of climate change, including ocean warming and enhanced storms, and which are also best able to repopulate neighbouring regions through dispersal of their planula larvae when, hopefully, the climate has stabilized. This research has now been completed with fifty regions identified that harbour coral reefs that have a relatively good chance of surviving the worst impacts of climate change. These regions include sites in the Philippines, Borneo, Indonesia, the Great Barrier Reef, French Polynesia, East Africa, the Red Sea, and the Caribbean. The next step is to develop and implement management plans that will protect the coral reefs at these sites from local impacts, such as overfishing and pollution, to enhance their resilience to the stresses of climate change.

Another initiative is to identify coral reef 'oases' around the planet where small patches of coral appear to be surviving, and even thriving, against the worst effects of climate change and local impacts, including bleaching, storms, and crown-of-thorns outbreaks. Many such oases have so far been identified in the Pacific and Caribbean. In some instances, the oases appear to have so far escaped the worst impacts on account of their location, for example, away from major storm tracks, in deeper water less affected by storms, or experiencing low levels of pollution. In other cases, the corals appear to have biological traits that have adapted them to better resist the effects of high ocean temperatures, or the oases systems have ecological characteristics, such as adequate numbers of herbivores, that allow them to recover more quickly from bleaching events or crown-of-thorns damage. These small oases are helping identify areas that can be prioritized for conservation and may, in future, act as sites from which to repopulate neighbouring regions, either through passive larval dispersal or by actively outplanting corals with favourable characteristics to other sites.

Many coral reef scientists and managers now believe that such measures alone will not be enough to preserve and restore coral reefs and the ecological services they provide for future generations. They argue that more active interventions involving new approaches and technologies are urgently required. One suggested approach involves 'assisted gene flow' in which corals already resistant to higher ocean temperatures are relocated to reefs in cooler waters where they will have a better chance of survival as ocean temperatures increase there. This could involve, for example, transferring heat adapted corals from the northern Great Barrier Reef to cooler reefs in the southern Great Barrier Reef. Another approach, termed 'assisted evolution', is being developed by some research groups. One example of assisted evolution is to develop strains of coral species that are adapted to higher temperatures using selective breeding techniques. These thermally tolerant corals could then be farmed and outplanted to restore reefs. It has also been proposed that modern gene editing techniques, such as the use of CRISPR-Cas9, should be used to speed up assisted evolution to engineer superior strains of corals, and such techniques are already being trialled.

Although the widespread adoption of such radical technologies may help to preserve and restore coral reefs in 'high value' locations it would seem unlikely that they could cost-effectively be used to resurrect vast areas of reef ecosystems to their historical pre-disturbance state. In this case, artificial 'engineered' reef habitats may need to be developed that replicate some of the ecosystem services provided by natural reefs such as habitat for fisheries and coastal protection. Clearly, coral reef biology is, out of necessity, on the cusp of a new era focused on developing new approaches for protecting and restoring coral reef ecosystems and preserving at least some of their beauty and ecosystem services in the face of rapid climate change.

## Mangroves

Mangrove is a collective term applied to a diverse group of trees and shrubs that colonize protected muddy intertidal areas in tropical and subtropical regions, creating mangrove forests, or mangals.

Two of the most common species of mangrove trees are the red mangrove, *Rhizophora mangle*, and the black mangrove, *Avicennia germinans*. The red mangrove can be distinguished by its characteristic tangle of prop roots that help support the tree in soft sediment. The black mangrove has a more typical tree-like trunk surrounded by a mass of peg-like structures, called pneumatophores (see Figure 28), which arise from a system of roots buried in the sediment and radiating out from the trunk.

Mangroves inhabit a very harsh environment consisting of sediments saturated with seawater and full of decaying matter,

28. **Black mangrove tree,** *Avicennia germinans*, **St Thomas, US Virgin Islands, showing an extensive system of pneumatophores (P).**

which makes them anaerobic, or lacking in oxygen. To deal with
the lack of oxygen, mangrove trees have evolved root systems that
are adept at extracting oxygen from the air or from the seawater
when they are submerged. The prop roots of the red mangrove
are covered with small nodular structures called lenticels through
which oxygen is supplied to the underground root system. The
peg-like pneumatophores of the black mangrove, which are also
covered with lenticels, act like snorkels, drawing oxygen from
the air or surrounding seawater (see Figure 28).

Mangrove trees have also adapted to growing in salty sediments.
They deal with excess salt in several ways. Mangrove tree roots
and stems have special tissues which act as a barrier to reduce
the amount of salt that enters the plant. Nonetheless, some salt
does penetrate the plant, which can tolerate salt concentrations in
its sap that are ten to a hundred times greater than is found in
normal plants. In addition, the leaves of the black mangrove
possess special glands that concentrate and excrete excess salt.
The salt crystals collect on the under surface of the leaves and are
washed off during rain. Mangroves also concentrate salt in old
leaves, bark, flowers, and fruit, which take away the salt when they
drop off the tree.

Mangrove trees produce flowers pollinated by the wind or by bees.
The flowers produce seeds which germinate into seedlings which
grow into young trees with distinctive cigar-shaped stems while still
on the tree. These young trees, or propagules, are ready to produce
roots as soon as they fall from the parent tree and encounter a
suitable habitat. The propagules are buoyant and can survive in
seawater and can drift with the currents for more than a year.
Once stranded on a suitable shore they quickly take root and grow.

Mangroves form a complex and productive habitat. Few organisms,
except for a few crabs, can graze directly on the mangrove leaves.
But mangroves are constantly dropping off dead leaves and
branches, which are broken down by bacteria and fungi, and form

the basis of a productive food web. Crabs and shrimps and other organisms graze on this detrital material and are then eaten by fish, turtles, and shorebirds.

Mangroves are of great importance from a human perspective. The sheltered waters of a mangrove forest provide an important nursery habitat for many species of juvenile coral reef fishes and for crabs and shrimp. Many commercial fisheries depend on the existence of healthy mangrove forests, including blue crab, shrimp, spiny lobster, and mullet fisheries. Mangrove forests also stabilize the foreshore and protect the adjacent land from erosion, particularly from the effects of large storms and tsunamis. For example, studies have shown that the area of flooding in south-western Florida caused by Hurricane Wilma in 2005 would have extended 70 per cent further inland if not for the protection provided by a zone of coastal mangroves. Mangroves also act as biological filters by removing excess nutrients and trapping sediment from land run-off before it enters the coastal environment, thereby protecting other habitats such as seagrass meadows and coral reefs.

Large-scale destruction of mangroves occurs naturally as a result of hurricanes, typhoons, and cyclones which uproot trees, or smother the roots with excess sediment. Mangroves can generally recover from such events in two or three decades. Unfortunately, most of the destruction of mangroves is now brought about by human activity. Mangrove trees are heavily harvested by humans for timber, firewood, and to produce charcoal. They are also routinely cleared to make way for coastal development. Furthermore, mangroves are frequently converted into agricultural land, for example for rice farming, and into large ponds for the culture of shrimp and fish and to produce salt.

As a result of this heavy human pressure, mangrove forests are disappearing rapidly. In a twenty-year period between 1980 and 2000 the area of mangrove forest globally declined from around

20 million hectares to below 15 million hectares and mangroves currently occupy just over 8 million hectares. In some specific regions the rate of mangrove loss has been truly alarming. For example, Puerto Rico lost about 89 per cent of its mangrove forests between 1930 and 1985, while the southern part of India lost about 96 per cent of its mangroves between 1911 and 1989. Concerted conservation and management efforts are required at community and local and national government levels to stem the pace of mangrove destruction in the face of a rapidly growing human population. As much as possible, existing stands of healthy mangroves need outright protection. Additionally, policies and strategies need to be developed and implemented to begin to regenerate degraded mangrove habitats in critical areas to restore their ecosystem services.

# Chapter 6
# Deep-ocean biology

There is no single definition of the deep ocean. Traditionally, the 200-metre depth was used to define the boundary between 'shallow' and 'deep' ocean, but other schemes have used 800, 1,000, or even 2,000 metres as the boundary. Here we will follow the current definition of the World Register of Deep-Sea Species (WoRDSS) that defines the deep ocean as the water column and ocean bottom greater than 500 metres, which is the depth at which seasonal variation in temperature and salinity and the influence of sunlight is minimal. This habitat encompasses roughly 90 per cent of the ocean's volume yet this vast region is the least investigated and understood environment on the planet. To put this into perspective, oceanographers have so far mapped about 10 to 15 per cent of the bottom of the Global Ocean at approximately 100-metre resolution, while astronomers have mapped practically the entire surface of Venus, Mars, and the moon at similar resolution, and much of Mars at about 20-metre resolution, and much of the moon at about 7-metre resolution. We therefore know more about the geography of objects in our solar system than we do about the largest habitat on our planet. Furthermore, at present only about 0.05 per cent of Earth's ocean floor has been mapped to the detail that allows us to detect features a few metres in size. Even less of the ocean floor has been observed from a crewed submersible or by remote operated vehicle. By comparison, 12 people have spent a total of 300 hours on the surface of the moon, whereas 8 people have spent less than 19 hours exploring the Challenger Deep, the deepest part of the Global Ocean.

## The physical environment of the deep ocean

The deep ocean, except near its upper boundary, is devoid of sunlight, the last remnants of which cannot penetrate much beyond 200 metres in most parts of the Global Ocean, and no further than 800 metres or so in even the clearest oceanic waters.

Extreme pressure is another defining characteristic of the deep ocean. Seawater is a heavy substance and a column of seawater 10 kilometres high—typical of the deeper parts of the Global Ocean—exerts a pressure of 10,000 tonnes per square metre, which is about the weight of fifty-five jumbo jets.

Except in a few very isolated places, the deep ocean is a permanently cold environment, with sea temperatures generally ranging from about 2°C to 4°C. Since the deep ocean is mostly below the oxygen minimum zone, usually present at depths from about 200 to 1,000 metres, dissolved oxygen concentrations are generally more than adequate to support life.

Food is scarce in the deep ocean. Since there is no sunlight, there is no primary production of organic matter by photosynthesis. The base of the food chain in the deep ocean consists mostly of a shower of 'marine snow'—particulate organic matter (POM) sinking slowly through the water column from the sunlit surface waters of the ocean. This is supplemented by the bodies of large fish and marine mammals that sink more rapidly to the bottom following death, and which provide sporadic feasts for deep-ocean bottom dwellers.

The picture of the deep ocean that thus emerges is one of a cold, dark, extremely pressurized and food-limited habitat—a very harsh and extreme environment from a human perspective. And yet, this immense habitat possesses a great diversity of marine life beautifully specialized for living under such conditions.

## Adaptations of deep-ocean animals

In the pelagic environment of the deep ocean, animals must be able to keep themselves within an appropriate depth range without wasting energy in their food-poor environment. This is often achieved by reducing the overall density of the animal to that of seawater so that it is neutrally buoyant. Thus, the tissues and bones of deep-ocean fish are often soft and watery. The deep-ocean pelagic environment is also dominated by gelatinous animals such as jellyfish, siphonophores, ctenophores, and salps whose body density is close to that of seawater.

Since food is scarce for deep-ocean fish, full advantage must be taken of every meal encountered. Compared to fish in the shallow ocean, many deep-ocean fish have very large mouths capable of opening very wide, and often equipped with numerous long, sharp, inward-pointing teeth. Good examples include the gulper eel, anglerfish, loosejaw, and black swallower (see Figure 29(a)). These fish can capture and swallow whole prey larger than themselves so as not to pass up a rare meal simply because of its size. These fish also have greatly extensible stomachs to accommodate such meals.

Although no sunlight penetrates the deep ocean, it sparkles with another kind of light that is of biological origin—the glow and flashes of bioluminescence. This is created by a chemical reaction in specialized organs called photophores present in the bodies of many deep-ocean animals including fish, octopuses, squid, jellyfish, worms, crustaceans, and sea stars. The animals generally produce the light themselves, although in some cases they have evolved a symbiotic relationship with bioluminescent bacteria which produce the light.

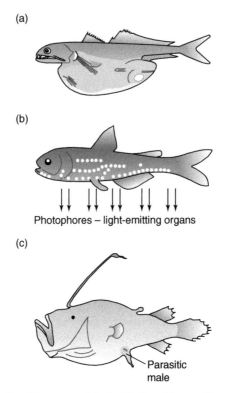

29. **Examples of deep-ocean fish: (a) Black swallower with an engulfed prey fish; (b) Lanternfish showing photophores along the side and belly; (c) Anglerfish with attached parasitic male.**

Bioluminescence is a major adaptation in the deep ocean and has evolved separately in many different animal groups to serve vital functions such as mating, finding food, and avoiding predators. For example, lanternfish possess rows of photophores along their belly and sides which emit blue, green, or yellow light (see Figure 29(b)). These are arranged in a pattern specific to each species of lanternfish and, in some species, the pattern differs between males and females. Their large eyes allow these fish to detect the light being emitted by other members of the same species which is used for communication and identifying a mate in the vastness of the deep ocean.

Some deep-ocean fish use bioluminescence to attract scarce prey. Anglerfish possess a long, flexible tentacle—a kind of fishing pole—that extends upwards between the eyes, at the end of which is a luminescent organ called the 'esca' (see Figure 29(c)). The esca is wiggled to resemble a small fish which acts as a lure to attract prey fish close enough to be engulfed whole. Contact with the esca automatically triggers the jaws to respond in trapdoor fashion. Variations on this theme are common in deep-ocean fish, many of which have light-emitting appendages located near their jaws which act as lures.

Many deep-ocean animals use bioluminescence to defend themselves against predators. For example, when the vampire squid is approached by a predatory fish it produces an array of flashes that startle the predator. If attacked further, it will release a fluid loaded with luminescent particles that envelopes the predator and helps the vampire squid to escape.

Finding a mate in the vastness of the deep ocean is a challenge. Anglerfish have solved this problem through an unusual adaptation—the males of most species have been reduced to a tiny parasitic form that attaches itself permanently to the much larger females (see Figure 29(c)). His mouth fuses to her body and his blood vessels merge with hers. Thus, a male is always present to fertilize the eggs of the female, eliminating the need for the female to find a male at breeding time. Of course, the dwarf male will have to locate a female in the first instance, presumably by smell or by being attracted to the bioluminescent esca, but once attached, the problem of finding a mate is solved.

Not surprisingly, deep-ocean animals have evolved adaptations for life in a highly pressurized environment. These high pressures not only impact on them structurally, but also have profound effects on their physiology and biochemistry. High pressure affects the physiology of cell membranes by compressing them, expelling fluid, and making them more rigid, and thus less capable of channelling nutrients and wastes in and out of the cell. Deep-ocean organisms have developed biochemical adaptations to maintain the functionality of their cell membranes under pressure, including adjusting the kinds of lipid molecules present in membranes to retain membrane fluidity under high pressure. High pressures also affect protein molecules, often preventing them from folding up into the correct shapes for them to function as efficient metabolic enzymes. In response, deep-ocean animals have evolved pressure-resistant variants of common enzymes that mitigate this problem.

## Mass migrations from the ocean depths

An extraordinary behaviour is exhibited by masses of many kinds of deep-ocean animals—a daily vertical journey from the depths into shallower water, perhaps the planet's largest animal migration. As sunset sweeps across the oceans, vast numbers of marine animals including copepods, shrimp, jellyfish, squid, salps, and fish swim up from depths of 1,000 metres or more towards the surface. Some 5 billion tonnes of animals are estimated to move towards the surface each night, including huge numbers of lanternfish, one of the most abundant mid-water fish in the oceans. At sunrise, this aggregation of pelagic animals sinks back down into the depths again. The mass of migrating animals is so dense that it reflects the sound waves from ships' sonar systems, showing up as a distinct mid-water 'false bottom' called the Deep Scattering Layer (DSL). The daily variation in light level seems to be the important cue for maintaining the migration, although vertical migration also occurs under the Arctic ice during the constant darkness of winter. An internal biological clock must therefore also be involved.

The participants in this mass migration must derive some considerable benefit from undertaking this huge journey every day. The most accepted explanation at this time is that the zooplankton, such as copepods, are migrating to the surface to feed on phytoplankton at night; presumably, the copepods are less visible in the dark to visual predators such as fish, which are more abundant in shallower water. They then leave the photic zone during the day to avoid being seen. The lanternfish and other small consumers are presumably undergoing the migration to also avoid predators and to follow their zooplankton food. The migration may also confer an energetic advantage on some of the participants because they are spending the day in deeper, colder water where metabolic rates are lower and they can conserve energy while digesting their night-time meal.

## Energy flow in the deep ocean

Almost all food for deep-ocean organisms comes from primary production in the photic zone. This food is in the form of dead organic material that sinks from the ocean surface as marine snow. This consists of small sticky clumps, or aggregates, of organic particles that include phytoplankton cells, dead zooplankton, and the faecal pellets produced by zooplankton. These aggregates sink slowly through the water column at a rate of about 100 to 200 metres per day. It thus takes in the order of weeks for them to reach the deep-ocean floor. Along the way, the nutritional value in the aggregates is being extracted by bacteria in the water column; thus, the deeper they sink, the more they become depleted of nutritious substances. When this organic material finally reaches the ocean floor, what food energy remains can be used by benthic animals. Some are suspension feeders—equipped to filter organic particles suspended in the layer of seawater just above the bottom. Others are deposit feeders—able to consume organic materials that have accumulated on or in the surface sediments. Predatory benthic

invertebrates feed on these particle feeders and bottom-dwelling demersal fish forage on the benthic fauna.

Because organic material accumulates on the bottom over very long periods, the deep-ocean sediments contain a considerable amount of organic matter in the form of DOM and POM which form the basis of a microbial food web, with abundant bacteria and archaea using this material as a food source. The sediments also contain extraordinary numbers of viruses—about 1 billion per gram of sediment. These create a viral loop by infecting the bacteria and archaea which release DOM and POM back into the sediments and water column when they die. It appears that much of this energy is rapidly recycled within the viral loop and not much is available for larger organisms such as protists and larger zooplankton.

## Diversity of deep-ocean benthic animals

The deep-ocean bottom was once considered to be a biological desert, but as more high-quality samples were obtained in the 1960s and more observations made from submersibles and through the cameras of remote-operated vehicles, it became clear that a spectacular diversity of animals inhabit the deep-ocean floor. These include many different species of suspension-feeding invertebrates such as sponges, sea lilies, sea pens, sea anemones, sea fans, and fan worms; deposit-feeding animals such as worms, sea cucumbers, brittle stars, sea urchins, and clams; and predators such as sea urchins, sea stars, sea anemones, amphipods, and octopuses (see Figure 30).

30. **Deep-ocean benthic diversity: (a) Sea cucumber at 2,660 metres depth. Note mouth (M) surrounded by a ring of feeding tentacles that pick-up food particles from the sediments; (b) Anemone-like animal clinging to a sponge stalk at approximately 4,000 metres depth; tentacles can be up to two metres long; (c) Octopus dubbed 'Casper' at 4,000 metres depth.**

# Hydrothermal vent and cold seep communities

The idea that all food in the deep ocean is derived from the surface is not strictly true. There are some very remarkable spots on the deep-ocean floor where food is created in place. This is a form of primary production that is not driven by the energy in sunlight but by the energy present in chemical compounds. Submarine hydrothermal vents represent one example of deep-ocean life driven by chemical energy.

Hydrothermal vents were discovered from the deep-diving submersible *Alvin* in 1977 on an ocean ridge near the Galapagos Islands at around 2,700 metres. Since then many other hydrothermal vents have been found throughout the Global Ocean, mostly on ocean ridges, and it has been estimated that many thousands of vents are active at any one time on the ocean floor.

Hydrothermal vents are created when seawater seeps deep down into the ocean floor where it reacts with hot rock to form a superheated fluid laden with chemicals. This fluid is ejected at high pressure back into the ocean through fissures as a hot water geyser. Generally, several geysers are clustered together to form vent fields that range from pool table to tennis court size.

Typically, hydrothermal vents eject black-coloured fluid through chimney-like structures that can be tens of metres in height (see Figure 31). The vent fluid from these 'black smokers' is highly acidic and darkened by tiny suspended sulphur-containing mineral particles that precipitate out of the vent fluid as it mixes with the surrounding cold seawater. The vent fluid from black smokers can exceed 400°C when it first exits the ocean floor. Other hydrothermal vent systems, called 'white smokers', form white chimneys formed of calcium carbonate that can be 60 metres tall and eject a highly alkaline lighter-coloured fluid at lower temperatures containing large quantities of hydrogen gas and methane.

**31. A mass of tube worms (left of the picture) bask in warm seawater near a black smoker spewing 400°C fluid at a depth of 2,250 metres.**

Many kinds of bacteria and archaea thrive near hydrothermal vents. They use chemical compounds present in vent fluids, including hydrogen sulphide, hydrogen, and methane, as an energy source to produce organic compounds such as glucose. The details of some of the many different chemosynthetic processes used by bacteria and archaea associated with hydrothermal vents are yet to be fully worked out. One widely used process, however, involves the breakdown of hydrogen sulphide ($H_2S$) to create energy to produce organic compounds from carbon dioxide that they absorb from the seawater. The process can be summarized with the following equation:

$$6O_2 + 6H_2S + 6CO_2 + 6H_2O \rightarrow C_6H_{12}O_6 + 6H_2SO_4$$

organic     sulphuric acid
compounds

This form of chemosynthesis requires oxygen to drive the process and is therefore known as aerobic chemosynthesis. It is ultimately linked to sunlight because photosynthesis is the source of oxygen on the planet. On the other hand, chemosynthetic microorganisms that use hydrogen or methane from white smokers as an energy source do not require oxygen, a process termed anaerobic chemosynthesis. Furthermore, the hydrogen and methane that they use for energy can be generated by purely geochemical processes taking place beneath the ocean floor. Thus, this form of chemosynthetic primary production supports an intriguing and what might seem a rather alien form of life that has no link to the sun and light-driven photosynthesis.

The chemosynthetic bacteria and archaea associated with vents support an astonishing animal community that includes giant clams, giant mussels, and various species of crabs, limpets, sea anemones, worms, shrimp, amphipods, stalked barnacles, octopuses, and fish. Most of the species found at vents were previously unknown before the vents were discovered and are found only at vent systems.

Some of the bacteria and archaea are free living, suspended in the vent plume or forming mats on the rocky bottom adjacent to the vent. Some vent-associated animals, such as clams and mussels, filter these microorganisms from the seawater while others, such as limpets, graze on the microbial mats. Most vent animals do not, however, feed directly on these free-living microorganisms. Instead they have developed complex symbiotic relationships with chemosynthetic bacteria from which they derive their nutrition. A good example is the giant tube worm, *Riftia pachyptila*, which is abundant at many vent sites in the Pacific Ocean (see Figure 31). These peculiar worms live within white protective tubes up to two or more metres in length and have no trace of a digestive system as adults. They possess a bright red plume that can be extended out of their tube to absorb hydrogen sulphide and oxygen from their surroundings. These chemicals are then transported within the blood system of the worm to a specialized stomach-like structure called the trophosome that is filled with chemosynthetic bacteria that produce food for themselves and their tube worm host. Other vent species have evolved other kinds of symbiotic relationships with chemosynthetic bacteria. For instance, Yeti crabs, which are 15-centimetre-long crabs not discovered until 2005 on vents south of Easter Island, feed on chemosynthetic bacteria that they cultivate on their body. The giant clams and mussels at vents foster chemosynthetic bacteria in the tissues of their gills. The organic matter produced by the bacteria provides most of the food for the clams and mussels, and although they are capable of filter-feeding, they will die without the nutrition produced by their symbiotic partners.

Hydrothermal vents are unstable and ephemeral features of the deep ocean. Repeated observations of known vent systems show that the rate of flow and chemical composition of the vent fluids can vary over a period of months, and dead vents surrounded by the remains of vent communities have been observed. The lifespan of a typical vent is probably in the order of a decade up to a century. Since many vent animals can live only near vents, and the distance between vent systems can be hundreds or thousands of kilometres, it is a puzzle how vent animals escape a dying vent and colonize other distant vents or newly created ones. It is known that some vent animals grow very fast and can reach a large body size and sexual maturity before the vent dies. For example, the giant tube worm can grow as much as two metres in a year. On account of their large size, many vent animals can produce large numbers of planktonic larvae that slow-moving deep-ocean currents may disperse over long distances, allowing them to colonize other vent sites. In some species, the larvae may rise to the surface where they are dispersed more rapidly by surface currents before sinking back to the ocean floor where they may encounter a suitable vent site to colonize.

Hydrothermal vents are not the only source of chemical-laden fluids supporting unique chemosynthetic-based communities in the deep ocean. Hydrogen sulphide and methane also ooze from the ocean bottom at many locations at similar temperatures to the surrounding seawater. Since 1983 many such 'cold seeps' have been discovered scattered along the continental margins of all the world's oceans at depths from the intertidal to over 7,600 metres. The communities associated with cold seeps are much like hydrothermal vent communities, with many of the animals having a close relationship with chemosynthetic bacteria. Clams, mussels, sponges, and crabs can be very abundant, along with dense thickets of tube worms. Cold seeps appear to be more permanent sources of fluid compared to the ephemeral nature of hot water vents. Thus, in contrast to hydrothermal vent communities, cold seep communities can possess slow-growing, long-lived species. For example, some tube worms associated with cold seeps are estimated to be 250 or more years old.

## Whale falls and other deep-ocean food riches

Although the deep ocean is one of the most food-limited environments on the planet, large packets of food do arrive on the bottom and create a localized nutritional oasis for a host of deep-ocean animals. The bodies of large marine animals that sink rapidly in an intact condition are the source of such food bonanzas.

'Whale falls', the carcasses of dead whales on the deep-ocean floor (see Figure 32), have been observed from submersibles and using side-scan sonar devices. It was once assumed that whale falls were rare events in the deep ocean but there is growing evidence that such feasts occur more frequently than thought. It has been estimated that at any one time there are hundreds of thousands of whale carcasses on the bottom of the Global Ocean in various stages of decomposition, each providing the nutritional value of roughly 2,000 years of marine snow accumulating at the site of the carcass. The average distance between whale falls on the ocean floor has been calculated to be about 12 kilometres and perhaps much less than this beneath the migration routes and breeding and feeding grounds of the more abundant whale

**32. Remains of a whale fall in Monterey Canyon off the coast of California. This photo was obtained by MBARI researchers Robert Vrijenhoek and Shana Goffredi during a February 2002 dive using MBARI's remotely operated vehicle, *Tiburon*. The (red) 'fuzz' on the whale bones represents thousands of deep-ocean worms feeding on the fats and oils in the bones.**

species. It is quite likely that whale falls were much more common on the deep-ocean floor prior to the great reduction in whale numbers resulting from human harvesting in the 1800 to 1960s period.

When these large 30–160-tonne corpses first arrive on the ocean floor they are located by smell within days by large aggregations of mobile scavengers such as hagfish, rattails, sharks, crabs, and amphipods which relentlessly remove much of the fleshy material over the course of months to a year, depending on the size of the whale. The whale bones, and also the sediments around the whale carcass, which have been enriched with material derived from decomposing flesh, are then colonized by huge numbers of worms, crustaceans, and other invertebrates that over the next couple of years consume the fats and oils present in the bones and sediments. These include specialized worms, called boneworms, which extend root-like structures into the whale bones. These 'roots' contain bacteria that help digest the fats and transfer the nutrients to the worms. In the final stage, which lasts for decades, chemosynthetic bacteria, and animals hosting such bacteria, such as clams and mussels, use hydrogen sulphide oozing from the decaying organic matter in the bones and sediments as an energy source. Interestingly, many of the animal species colonizing the final stage of whale fall decomposition are also found associated with hydrothermal vents and cold seeps. It is thus possible that the many whale falls scattered on the ocean floor act as shorter distance 'stepping stones' allowing these species to disperse more easily by means of their planktonic larval stages across the larger distances separating hydrothermal vent and cold seep communities.

## Seamounts—unique deep-ocean ecosystems

Seamounts are sites of significant biological activity in the deep ocean. They rise abruptly from the ocean floor and their peaks can be thousands of metres beneath the ocean surface. In contrast to the surrounding flat, soft-bottomed abyssal plains, seamounts provide a complex rocky platform that supports an abundance of organisms that are distinct from the surrounding deep-ocean benthos.

The summit and flanks of seamounts are frequently dominated by a dense thicket-like community comprising cold-water stony corals, soft corals, sea fans, black corals, and sponges (see Figure 33). Unlike shallow-water, reef-building corals of the tropics, which obtain much of their food from their photosynthetic zooxanthellae, deep-ocean stony corals lack zooxanthellae and rely solely on filtering zooplankton and suspended organic particles from the seawater. These stony corals grow very slowly and can be several hundred years old. Other types of corals associated with seamounts also show extreme longevity, with life spans of thousands of years. For example, a type of black coral, *Leiopathes*, sampled from a seamount in the Pacific Ocean, was shown using radiocarbon dating to be about 4,200 years old, making it among the world's longest-living colonial animals. These coral-dominated thickets create habitat for a host of other animals such as feather stars, sea lilies, brittle stars, sea stars, sea cucumbers, barnacles, sea squirts, worms, shrimp, and dense aggregations of fish.

33. **Feather stars and soft corals living at a depth of about 1,200 metres on Davidson Seamount off the coast of California. This photo was obtained using MBARI's remotely operated vehicle, *Tiburon*, during expeditions funded by the National Oceanic and Atmospheric Administration's Office of Exploration and the David and Lucile Packard Foundation (through MBARI).**

The high productivity of seamount communities is driven by several factors which vary with the depth of the seamount and the patterns of ocean circulation in its vicinity. Seamounts with peaks relatively close to the surface can obstruct the downward-migrating Deep Scattering Layer, trapping and concentrating large numbers of zooplankton each night. These provide a source of food for abundant suspension feeders and plankton-eating fish. Seamounts can also interact with and modify the horizontal currents flowing past them. This can result in the creation of 'Taylor columns', or vortices of rotating seawater that are retained over the summit of the seamount. These can further serve to retain and concentrate zooplankton around the summit and flanks of the seamount.

Seamounts support a great diversity of fish species, with around 800 species recorded living around seamounts. In the 1960s, deep-ocean fishing vessels looking for new stocks of fish began to trawl seamounts and discovered large aggregations of commercially important species. This triggered the creation of new deep-ocean fisheries focused on seamounts. Bottom trawls are towed from the summit down the flanks of seamounts to capture the fish. Commercial fish species that are targeted include orange roughy, oreos, alfonsino, grenadiers, and toothfish. These fish are not generally permanent residents of seamounts but aggregate at seamounts at certain times of the year to spawn, to feed on squid and small fish, or simply to rest. They are very slow-growing and long-lived and mature at a late age, and thus have a low reproductive potential. A good example of this is the orange roughy, which is known to live for more than 120 years and reaches maturity at around 30 years of age, with the females producing relatively small numbers of eggs. Such a life history is typical of many deep-ocean fish species.

Seamount fisheries have often been described as mining operations rather than sustainable fisheries. Fisheries typically collapse within a few years of the start of fishing and the trawlers then move on to other unexploited seamounts to maintain the fishery. The recovery of localized fisheries will inevitably be very slow because of the low reproductive potential of these deep-ocean fish species.

The destruction of fish stocks is not the only concern associated with seamount fishing. The trawling of seamounts causes extensive damage to the fragile coral communities, with the trawls bringing up not only fish, but large numbers of stony corals, black corals, and other benthic animals associated with the corals. The intensity of trawling on seamounts can be very high, with many hundreds to thousands of trawls often carried out on the same seamount. Tens of tons of coral can be brought up in a single trawl, and in one new seamount fishery it was estimated that almost one-third of the total catch consisted of coral bycatch. Comparisons of 'fished' and 'unfished' seamounts have clearly shown the extent of habitat damage and loss of species diversity brought about by trawl fishing, with the dense coral habitats reduced to rubble over much of the areas investigated.

Unfortunately, most seamounts exist in areas beyond national jurisdiction, which makes it very difficult to regulate fishing activities on them, although some efforts are under way to establish international treaties to better manage and protect seamount ecosystems. Not surprisingly, seamount-based fisheries are controversial. The New Zealand orange roughy fishery provides a good example of the history of seamount fisheries and current views regarding their sustainability. This fishery began in the late 1970s and expanded very rapidly, and by the 1980s annual catches had peaked at about 54,000 tonnes. By then it was clear that orange roughy stocks were much less productive than thought and that they were being severely overfished. The total allowable catch was reduced to 25,000 tonnes by the mid-1990s and to around 17,000 tonnes by 2000, and several stocks were completely closed to fishing. Many large food retailers banned the sale of orange roughy from their shelves and the fishery had basically collapsed. The New Zealand government together with the fishing industry then developed improved methods for estimating the productivity of orange roughy and some stocks have since begun to rebuild. Annual catches are now around 7,000 tonnes and the emphasis has switched to 'taking less and earning more'. In 2016 three orange roughy stocks achieved certification by the London-based Marine Stewardship Council (MSC) as sustainable based on the size of the stocks, the impacts of the fishery on the wider marine ecosystem, and the quality of fishery management. About two-thirds of the total fishery is now under certification. One condition was that fishing vessels must return to the same trawl tow sites each year to minimize seamount coral destruction, and the industry must develop a plan to increase understanding of fishing impacts on coral. Nevertheless, not all conservation groups agree with the MSC certification, including WWF and Greenpeace, arguing that it is still too early to call this fishery sustainable, and New Zealand Forest & Bird still list orange roughy as a 'don't eat' choice in their 2017 Best Fish Guide.

# Chapter 7
# Intertidal life

The intertidal region of the Global Ocean is a thin strip of shoreline lying between the high and low tide marks—it is completely submerged by seawater at the highest high tides and completely uncovered at the lowest low tides. The intertidal region is occupied almost exclusively by marine organisms that have adapted to live in a very stressful physical environment influenced by exposure to air, temperature extremes, wind, and the pounding of waves. Although this region represents just a small part of the Global Ocean, it is home to a diverse and interesting marine community that people can study and enjoy on a routine basis on account of its accessibility. It is also a place where people routinely harvest seafood, and it is prone to a wide range of human impacts, including overharvesting, oil spills, coastal development, and the footsteps of thousands of visitors.

## Tides

The regular rise and fall of the tides are the dominant feature of the intertidal region. The driving force behind these tides is the gravitational pull of the moon and the sun on the fluid mass of the Global Ocean. Since the moon is much closer to the Earth than the sun, it has a greater effect in creating tides than the sun.

The moon causes the oceans on the side of the Earth closest to the moon to bulge out slightly. Another bulge in the oceans occurs on the opposite side of the Earth because, in simple terms, the Earth is also being pulled towards the moon and away from the water on this far side. The Earth continues to rotate beneath these two bulges and thus, in theory, any one point on the planet will pass beneath two bulges each day, which explains why tides on a shoreline often occur twice a day, roughly twelve hours apart.

The pull of the sun modulates the influence of the moon. When the Earth, moon, and sun are all roughly positioned in a straight line (each month at the full and new moon), the pull of the sun is added to that of the moon and thus the tides are highest around this time—these are the 'spring' tides. When the Earth, moon, and sun are at roughly right angles to each other (at the first and last quarters of the moon), the pull of the sun subtracts from that of the moon and the tides are lowest at around this time—'neap' tides.

In this way the moon and sun establish the basic rhythm and height of the tides on our planet. These are modified greatly by the continental land masses, which obviously interfere with the ocean bulges, and by the shape of the ocean basins and the peculiarities of the local shoreline. The result is that the moon and sun set up a kind of basin-scale tidal sloshing of the oceans that is modified regionally and locally to create different tidal patterns and heights at any one shoreline location. Thus, although most coastal locations have two low and two high tides per day of about the same height (termed semidiurnal tides), some have two high and two low tides of quite different heights each day (mixed semidiurnal tides), and a few places have only one high and one low tide per day (diurnal tides).

## Adaptations of intertidal organisms

The tides have a profound impact on intertidal marine organisms, which are regularly submerged by the incoming tide and then exposed for varying periods of time to air, heat, cold, rain, and waves on the receding tide. Intertidal organisms have evolved various ways to deal with such stresses. For example, small marine snails, or periwinkles, which live in the rocky intertidal zone of the tropics, have various ways to avoid overheating when the tide is out. They have light-coloured shells to reduce heat absorption, little bumps on their shells that act like the cooling fins on a radiator, and they hang on to the rocks as much as possible with a mucous thread to avoid direct contact with the hot substratum. Mussels and barnacles living in the rocky intertidal avoid water loss and desiccation at low tide by tightly closing their shells and trapping enough water inside to survive until the next high tide. Crabs seek shelter in crevices or under moist mats of seaweed, or simply retreat down the shore with the receding tide. Some seaweeds living in the intertidal can tolerate extremes of dehydration, losing up to 90 per cent of the water in their tissues at low tide. Others secrete a gelatinous mucous covering that helps lock in water. Intertidal animals and plants living in cold climates must cope with extremes of temperature, sometimes tens of degrees below freezing, for periods of hours or days when exposed to the air. Some of these organisms produce antifreeze compounds that help prevent their tissues from freezing or are otherwise extremely tolerant to becoming frozen.

Intertidal organisms on exposed shorelines must also deal with the crushing and dragging forces of waves. Adult barnacles and oysters deal with these forces by cementing themselves permanently to rocks, while limpets, snails, and chitons hang on tightly with muscular foot-like attachment structures; adult mussels secure themselves to rocks by secreting strong, thread-like fibres called 'byssal threads'. Intertidal algae use holdfasts to attach to a hard surface and have flexible fronds that can bend and twist with the waves without damage.

## Intertidal zonation

A classic feature of the intertidal region, particularly on rocky shores, is vertical zonation—the separation of intertidal life into

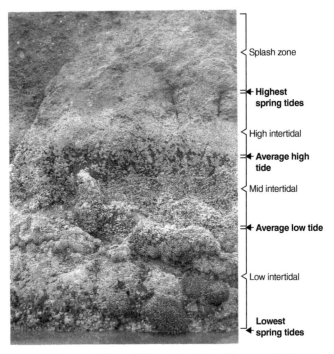

Splash zone

◀ **Highest spring tides**

High intertidal

◀ **Average high tide**

Mid intertidal

◀ **Average low tide**

Low intertidal

◀ **Lowest spring tides**

**34. Typical pattern of intertidal zonation at low tide on a rocky shore in Washington State, USA.**

prominent horizontal bands, often distinguished by different colours (see Figure 34).

The observed pattern of vertical zonation on rocky shores is quite similar from one region to another around the planet, which has led to a generally accepted 'universal' system for describing these zones. In this system the intertidal region is divided into four zones which, from highest to lowest on the shore, can be referred to as the splash zone, the high intertidal zone, the mid-intertidal zone, and the low intertidal zone (see Figures 34 and 35).

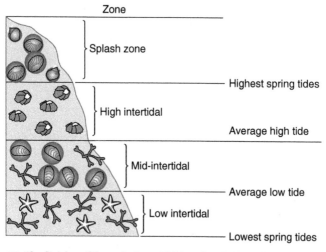

| | Zone |
|---|---|
| Splash zone | |
| | Highest spring tides |
| High intertidal | |
| | Average high tide |
| Mid-intertidal | |
| | Average low tide |
| Low intertidal | |
| | Lowest spring tides |

**35. The division of the rocky intertidal into four 'universal' zones.**

The splash zone is above the line of highest spring tides and its only direct connection with the marine environment is the spray from waves. This is a sparsely populated part of the shore because few organisms can withstand its extremely harsh conditions. The high intertidal zone is completely submerged only during spring tides, with parts exposed to the air for days to weeks, so it too is quite a harsh environment. The mid-intertidal zone is the area of the shore between about the average high tide and the average low tide levels, and thus most of it is submerged for prolonged periods during most tide cycles. This zone is densely covered by a variety of marine plants and animals. The low intertidal zone is the area between the average low tide level and lowest spring tides. Thus, it stays completely submerged during most tide cycles and is the least physically stressed part of the intertidal region.

On rocky intertidal shores in temperate regions the splash zone is colonized by patches of bright orange- and greyish-coloured lichens; blue-green cyanobacteria that form a thin black layer on the rocks; and hair-like green algae. Periwinkles and limpets are common in the splash zone, grazing on the bacteria and algae, along with isopods scavenging on dead organic material. For these animals, the splash zone is a refuge from predatory crabs and snails that cannot venture for long into this harsh environment.

The high intertidal zone is dominated by barnacles which are often so densely packed together that they create a distinct white band on the shore. Barnacles have a tiny, shrimp-like larval stage, called a cyprid, that drifts and swims about freely in the ocean. When the larva encounters a suitable place to settle in the intertidal region, it secretes glue from a gland in its head and cements itself onto a solid surface. It then secretes a calcareous box-like shelter around itself that is capped by two pairs of plates that can be opened and closed like a trapdoor. When the plates are opened at high tide, the barnacle extends a set of feathery legs that filter plankton out of the seawater. At low tide the plates are sealed shut to protect the animal inside from predators and from drying out.

The mid-intertidal zone is heavily populated by mussels which often create a distinct black band on the shore. Like barnacles, mussels also produce tiny free-living larvae, called veligers, that disperse for a period in the ocean and then seek a suitable place to settle in the intertidal region. When a mussel larva detects a good spot, it secures itself to the rock by secreting byssal threads from a gland in its foot and then grows and transforms into the familiar two-shelled adult. The adults filter plankton from the seawater when submerged and otherwise close their shells tightly to avoid dehydration and roaming predators such as sea stars. The mid-intertidal zone is also occupied by oysters, limpets, and periwinkles as well as various species of fleshy brown seaweeds which provide a moist shelter for sea stars, sea urchins, and other marine animals at low tide.

The low intertidal zone is richly populated by a range of seaweeds and animals that can tolerate occasional exposure to air. Here, red, green, and brown seaweeds proliferate, along with many kinds of marine animals including sea anemones, sea stars, sea urchins, brittle stars, sea cucumbers, crabs, snails, sea slugs, and worms.

On rocky intertidal shores in the tropics, various species of bacteria and algae create grey- and black-coloured films on the rocks in the splash zone. Here, a variety of periwinkles graze on these films. Rock-boring algae often give the high intertidal zone of the tropics a yellow appearance. Barnacles, limpets, and snails can be present in the high intertidal, but generally in small numbers compared to temperate rocky shores. A pink zone is often present in the mid-intertidal zone created by encrusting coralline algae. Various snails occupy this zone along with mussels, sea anemones, limpets, and barnacles. Brown seaweeds often cover rocks in the low intertidal zone, which is home to a variety of marine organisms such as sea urchins, sea anemones, limpets, sea cucumbers, and sponges.

## Causes of intertidal zonation

For many decades marine biologists have worked at understanding the factors that create the distinct patterns of vertical zonation on rocky intertidal shores. It is now known that intertidal zonation is caused by a complex interaction of biological and physical factors. In general, a seaweed or animal's tolerance to physical factors, such as exposure to air, heat, cold, desiccation, and wave forces, determines where in the intertidal region it can potentially live. But biological factors, such as competition, grazing, predation, and patterns of larval settlement, interact and modify the influence of physical factors, and ultimately determine where an organism actually occurs on the shore.

The biological interactions between barnacles, mussels, sea stars, and seaweeds on a temperate rocky shore provide a good example of how this works. Living space is at a premium in the intertidal region and hence competition for space is an important factor influencing the actual distribution of some species on the shore. Barnacles can settle and live anywhere in the mid- and high intertidal zones of temperate rocky shores because of their tolerance to physical stresses. However, they are generally excluded from much of the more benign mid-intertidal zone where they would be subject to less stress, have more time to feed, and grow faster. This is because mussels outcompete barnacles for space in the mid-intertidal. They do this by overgrowing and smothering the barnacles. Thus, barnacles ultimately persist mainly in the refuge of the high intertidal zone where mussels cannot tolerate the harsh environment.

Yet barnacles are not entirely excluded from the mid-intertidal and persist in patches here and there in this zone. This is a result of another biological interaction—predation. In the mid-intertidal, mussels can be eaten by sea stars and are thus prevented from completely dominating this zone. But sea stars can only venture up into the mid-intertidal to eat mussels for limited periods at high tide, and thus are prevented from overeating mussels. In the low intertidal zone, though, sea stars can prey on mussels mostly at will and so almost completely eliminate mussels in this zone.

Hence, in this situation, the upper limit of barnacle distribution is set by physical factors and the lower limit by a biological factor—competition for space from mussels; similarly, the upper limit of mussel distribution is set by physical factors and the lower limit by a biological factor, in this case predation by sea stars. The more closely the rocky intertidal region is studied, the greater the complexity of biological interactions revealed. For example, brown seaweeds are also competing for space in the mid-intertidal and, if they become well established, they can maintain their foothold by preventing mussel and barnacle larvae from settling there. This appears to be a result of the blades of the seaweeds being swept back and forth across the rocky surface by wave action and preventing mussel and barnacle larvae from securing an attachment. On the other hand, grazing on seaweeds by limpets and snails can reduce seaweed cover, allowing mussels and barnacles to become established. In response to grazing, some seaweeds can produce toxic chemicals that deter further grazing.

## Human impacts

The intertidal region can be subject to severe human impacts on account of its accessibility and very close association with land and human development. It supplies humans with a ready source of wild marine food such as mussels, oysters, limpets, periwinkles, sea urchins, crabs, clams, abalone, and various seaweeds which are often harvested on a recreational and largely unregulated basis. In some places close to populated regions, such organisms have been overharvested and are now scarce or absent in the intertidal where they were once abundant. Such harvesting can also fundamentally change the structure of intertidal communities. A good example is provided by the harvesting of a large predatory intertidal snail by local people in Chile. In areas where this snail is harvested, the mid-intertidal zone is dominated by a monoculture of mussels, but where such harvesting is prevented, the mid-intertidal is populated by barnacles and seaweeds, as well as mussels, and has a higher overall species diversity. This is because the snail eats mussels and so prevents the mussels from completely outcompeting other species for space.

In some jurisdictions around the world, steps have been put in place to more carefully manage the impacts of recreational harvesting in the intertidal by enforcing recreational quotas or by establishing 'no take' marine reserves where harvesting of any kind is not allowed.

Larger-scale commercial harvesting of intertidal organisms also occurs. For instance, an abundant brown seaweed, known as 'rockweed', is harvested on a commercial basis from intertidal regions of eastern Canada and Maine. The seaweed is dried and used as an organic fertilizer, animal feed, food supplement, and for the extraction of alginates which have many uses, including as an additive to ice creams and other dairy products. Wild mussels are also harvested commercially off the coast of Maine, while oysters are harvested commercially in the intertidal regions of South Africa and elsewhere around the world.

Just the simple act of walking and turning over rocks in the intertidal to observe the organisms underneath can be destructive. This is not an issue on isolated coasts, but rocky shores near urban centres attract huge numbers of visitors and can be highly impacted. To put this into perspective, some popular 'high-use' rocky intertidal sites along the Californian coast attract 25,000 to 50,000 visitors per year per 100 metres of shoreline. Less popular sites are still subjected to 2,000 to 10,000 visitors per year per 100 metres of shoreline. The trampling caused by all these visitors walking in the intertidal at low tide dislodges and crushes seaweeds and animals. Many visitors also turn over rocks, which will crush the organisms living on the top of the rock and expose those organisms living under the rocks to desiccation, wave action, and predation. This creates rocks which have only a fringe of organisms around their edges and none on the top and bottom surfaces. Even in no-take marine reserves, people can walk freely in the intertidal and so this kind of impact will still occur. Thus, if there is a desire to more fully protect some high-use intertidal regions, it is necessary to educate the public about the impacts they can cause and, in some instances, limit the number of visitors, keep them to intertidal 'trails', or restrict access completely at some sites.

In our fossil-fuelled civilization, oil pollution is a threat to intertidal communities just about anywhere on the planet, as evidenced by the ubiquitous tar balls—balls of crude oil—found washed up on beaches. Crude oil gets into the ocean from four major sources: natural seeps, extraction processes, transportation, and consumption. Based on data published in 2003 by the US National Research Council of the National Academy of Sciences, the breakdown of these sources is roughly as follows:

Around 600,000 tonnes of crude oil seeps naturally into the marine environment each year from oil-containing geological formations below the ocean floor. This represents almost half of all the crude oil entering the oceans. This large amount of crude oil is released at a slow enough rate, and from so many different locations, that the marine environment as a whole is not damaged.

The human activities associated with exploring for and producing oil result in the release on average of an estimated 38,000 tonnes of crude oil into the oceans each year, which is about 6 per cent of the total anthropogenic input of oil into the oceans worldwide. Although small in comparison to natural seepage, crude oil pollution from this source can cause serious damage to coastal ecosystems because it is released near the coast and sometimes in very large, concentrated amounts. In fact, the catastrophic Gulf of Mexico oil spill of 2010 is so far the largest accidental release of oil into the marine environment in history. In this incident, a deep-ocean drilling platform about 66 kilometres off the coast in about 1,500 metres of water failed, allowing huge amounts of crude oil to gush from a reservoir under the ocean floor. During a three-month period, about 670,000 tonnes of oil escaped into the Gulf of Mexico before the undersea gusher was finally capped. Around 790 kilometres of coastline were contaminated by oil which coated, smothered, and poisoned intertidal and subtidal marine life as well as coastal wildlife. It also resulted in the closure of shrimp fisheries in much of the Gulf. In addition, the chemical dispersants and mechanical devices used in the clean-up phase caused further damage to marine life.

The transport of oil and oil products around the globe in tankers results in the release of about 150,000 tonnes of oil worldwide each year on average, or about 22 per cent of the total anthropogenic input. Oil spills from tankers can be catastrophically large and concentrated, and thus very damaging. When the tanker *Exxon Valdez* grounded on rocks off the coast of Alaska in 1989, about 37,000 tonnes of oil were spilled into Prince William Sound. Currents carried the oil slick down the Alaskan coast, where it coated around 2,100 kilometres of coastline and extensively damaged the intertidal and killed hundreds of thousands of seabirds, thousands of marine mammals, and an indeterminate number of fish; also, fisheries for salmon, herring, crab, shrimp, rockfish, and sablefish were closed. The impacts of this disaster linger to this day with parts of the shoreline still contaminated with oil just beneath the surface. It is encouraging that the number of large (greater than 700 tonnes) tanker spills, and thus the total amount of oil entering the marine environment from this source, has decreased significantly over the last three decades, from an average of 7.7 spills per year between 1990 and 1999, to 3.2 between 2000 and 2009 and 1.8 between 2010 and 2017.

About 480,000 tonnes of oil make their way into the marine environment each year worldwide from leakage associated with the consumption of oil-derived products in cars and trucks, and to a lesser extent in boats. Oil lost from the operation of cars and trucks collects on paved urban areas from where it is washed off into streams and rivers, and from there into the oceans. Surprisingly, this represents the most significant source of human-derived oil pollution into the marine environment—about 72 per cent of the total. Because it is a very diffuse source of pollution, it is the most difficult to control. The impact of this chronic, diffuse oil pollution on marine organisms and the functioning of marine communities is insidious. It is known that the many kinds of organic hydrocarbons found in oil have deleterious and cumulative effects on many marine organisms, even at very low concentrations, with the larval stages of marine animals being particularly vulnerable.

# Chapter 8
# Food from the oceans

Humans have been harvesting food from the oceans for millennia. This would have begun as small-scale subsistence fishing in coastal waters but, with more experience and as population numbers increased, it evolved into a wide-ranging commercial enterprise. Now fishing the oceans takes place globally on an industrial scale and provides the human population with its last significant source of wild food. Today, there are more than 4.6 million fishing vessels on the oceans, around 2.8 million of which are motorized, and more than 40 million people work in the marine fishing primary sector. The landed value of wild-caught seafood is in excess of US$130 billion annually and seafood is a sought-after component of the diet of many billions of people worldwide, providing them with a source of high-quality protein. Unfortunately, global seafood catches peaked in the mid-1990s and are now stagnant or in decline, and fully a third of all fish stocks are overfished and facing collapse. Overfishing is seriously impacting not only the target species but also the biodiversity and functioning of many marine ecosystems.

## Historical expansion of seafood harvesting

It is often assumed that in ancient times catches of seafood were limited by low population numbers and the simplicity of fishing gear and vessels available, and that seafood stocks were only lightly exploited until recent times. On the contrary, it has been deduced from historical marine ecological studies that humans have had significant impacts on marine resources for thousands of years. The Mediterranean Sea provides a good example of long-standing human pressure on marine resources.

Humans have occupied the shores of the Mediterranean Sea on a sustained basis for about 50,000 years. Fishing communities have existed there for at least 10,000 years and seafood has been an important source of protein for Mediterranean people since at least the Greek and Roman periods, starting about 900 BC. Species harvested included dolphins, sea turtles, sharks, rays, tuna, sardines, anchovies, mullet, grouper, flatfish, oysters, mussels, clams, and scallops. Some of these species were depleted by Roman times and by the 1st century AD the coastal waters around Italy had been largely overfished and harvesting had spread to offshore islands like Sicily and Corsica. The collapse of the Roman Empire most likely relieved some pressure on marine resources but strong population growth recommencing in the 15th century resulted in renewed depletion of marine resources in coastal areas. In the late 19th century, fishing capacity grew greatly in the Mediterranean as population numbers increased exponentially. Through the early and middle 20th century, the fishing fleet became motorized and industrial fishing expanded into all offshore waters of the Mediterranean Sea. Now, after 100 years of intense fishing pressure, virtually all traditional marine food species in the Mediterranean Sea have been reduced to less than 50 per cent of their original abundance and about a third are now very rare and many are functionally extinct. The greatest impact has been on the top predators, which were the first to be targeted by fishers.

The intensive exploitation of marine resources got under way in other parts of Europe much later than in the Mediterranean. By about the 8th or 9th centuries, though, the Vikings were exploiting large stocks of cod, haddock, pollock, herring, and other species that flourished in the seas of northern Europe at that time. They brought this skill with them when they spread into Britain and Normandy, where marine fishing became well established by the 11th century. By the end of the 18th century, many of the fish stocks of northern Europe were in serious decline from overfishing and the first steps were taken to protect some stocks. Nonetheless, intensive fishing continued unabated and by 2000 more than half of northern European fish stocks were considered seriously overexploited. It is estimated that, collectively, the European fish stocks of today are just one-tenth of their size in 1900 (see Figure 36). Going back to a time when European fish stocks were unexploited, it has been suggested that at present European seas contain less than 5 per cent of the total mass of fish that once swam there.

Exploitation of marine food resources spread outwards from Europe surprisingly early in history. As early as AD 1000, Vikings were exploiting the rich fish stocks around Iceland and Greenland and possibly into the seas of northern Canada. Some 200 years later, Basque fishers were also harvesting North Atlantic fish

36. Scene at Grimsby fish market, UK in 1906. The amount and size of fish are extraordinary by today's standards.

stocks and may have been fishing off the coast of North America before Columbus' voyage of 1492. Nonetheless, by the early 1500s, Basque, French, and Portuguese fishers were routinely exploiting cod from the shallow Grand Banks off the coast of Newfoundland for the European market. By 1600, more than 150 shiploads of cod per year were being extracted from Canadian seas. The catch was preserved by either drying or salting it, in which form it could be stored for several years and provided a ready source of protein for Europeans. The permanent colonization of North America starting in the 17th century marked the beginning of the exploitation of the fish stocks off the east coast of the United States, including sturgeon, shad, salmon, alewife, and oysters. By the early 1800s, the fish stocks of the eastern seaboard of North America were showing signs of serious decline.

In the late 1800s and early 1900s, rapidly increasing urbanization and global population growth created an ever-increasing demand for seafood. This was linked to improvements in transport and preservation of the catch that meant that offshore fish stocks could be more easily exploited and distant markets served. The marine harvest increased at an unprecedented rate in size and geographic footprint from just after the Second World War, with a vast expansion of the global fishing fleet and improvements in fishing gear and the introduction of onboard processing of the catch. This marked the start of the mass industrialization of extraction of food from the sea. Fishing increased greatly in coastal waters throughout much of the world and fishing fleets began to exploit marine resources in deeper waters down the continental slopes, trawl seamounts, and pursue open-ocean species such as tuna across entire oceans. By 2000 fishing had intensified in just about all regions around the world and particularly in the Asian region, where China now reports by far the world's largest total marine catches.

From a 21st-century perspective it is almost impossible to conceive just how abundant fish once were in the oceans. However, the accounts of early explorers and settlers, along with articles and photographs from old newspapers and magazines, 19th-century fisheries statistics, and historical records of fishing competitions, paint a picture of astonishing past abundance and fish size compared to the present.

## Commercial fishing methods and their effects

The advent of motorized fishing vessels enabled super-efficient seafood catching with the use of trawls. Trawls are large, cone-shaped nets that are towed behind a fishing vessel, scooping up fish and other marine organisms. The mouth of a large trawl net can be about as wide as a football field and as high as a three-storey building.

Bottom trawling involves dragging a trawl along the ocean floor to catch demersal, or bottom-living, fish and invertebrates (see Figure 37(a)). Mid-water, or pelagic, trawls are towed off the bottom to catch schools of fish living in the water column. Commercial trawlers ply the shallow waters of the continental shelves and fish the continental slopes as well as canyons and seamounts down to depths of over 2,000 metres (see Figure 37(b)).

In heavily fished regions, bottom trawlers can fish the same parts of the ocean floor many times per year. Such intensive trawling causes great cumulative damage to the ocean floor. The trawls scrape and pulverize rich and complex bottom habitats built up over centuries by living organisms such as tube worms, cold-water corals, and oysters. These habitats are eventually reduced to uniform stretches of rubble and sand. These areas become permanently altered and occupied by a much changed and much less rich community adapted to frequent disturbance. Bottom trawls can also catch large amounts of unwanted, non-target species of fish and invertebrates, known as bycatch, which are often discarded. Furthermore, they can also unintentionally catch

Marine Biology

37. Bottom trawling: (a) Bottom trawl net being dragged along ocean floor; (b) Fishers unloading a trawl net full of orange roughy.

turtles, dolphins, and seals which drown in the nets. For these reasons bottom trawling is a particularly destructive form of fishing and has been banned from use in large areas of the oceans worldwide. Pelagic trawling is less destructive because there is no contact with the bottom. However, bycatch can still be a problem.

Another common harvesting method is longline fishing, which involves a vessel trailing a line from which baited hooks are suspended. There can be over 2,000 hooks hung from a line many kilometres long. Longlines can be set for catching pelagic or demersal fish, depending on the target species. This fishing method catches non-target fish species, as well as marine mammals, turtles, and seabirds. These issues can be mitigated by setting longlines at night to avoid catching seabirds which hunt visually, using weighted longlines that sink quickly, and using speciality hooks that do not easily snare non-target species.

Purse seines are used in the open ocean to catch fish such as tuna, sardines, salmon, and mackerel. Here, a school of fish is surrounded by a vertical curtain of netting, the bottom of which is drawn together to enclose the fish. Purse seines do not impact the ocean bottom, but they can capture non-target species, juvenile fish, and marine mammals, such as dolphins.

Gill net fishing involves hanging a curtain of netting to entangle fish. There is generally no contact with the bottom and the mesh size can be adjusted to minimize catch of juveniles. Nevertheless, gill nets carry the risk of bycatch and capturing protected species such as dolphins. Nets can be equipped with noise-producing 'pingers' that keep dolphins away.

## Commercially exploited marine species

There are thousands of species of marine fish but most of the global catch comprises a relatively small number of species falling into several main groups (see Figure 38). The clupeoid fish

Clupeoid fish (anchovy)

Gadoid fish (cod)

Flatfish (plaice)

Scombroid fish (tuna)

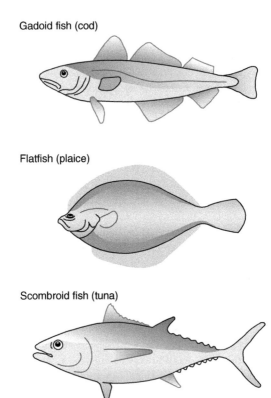

38. **Major groups of commercial marine species.**

account for a large proportion of the marine catch. These are small, schooling fish that feed directly on phytoplankton and zooplankton and include the herrings, sardines, and anchovies. The gadoid fish are another important group of commercial species. These are demersal fish and include cod, haddock, hake, and pollock, which are species that live in coastal waters of the North Pacific and North Atlantic Oceans. Flatfish such as flounder, halibut, sole, and plaice are another significant group of commercial bottom-dwelling species, whose habitat is also in coastal waters. Then there are the large, fast-swimming, pelagic carnivores, the scombroids, which include mackerels and tunas, the latter a major open-ocean fishery. Sharks comprise another exploited group of marine fish. There are also large fisheries for marine invertebrates, including crustaceans such as lobsters, crabs, and shrimp; and molluscs such as squid, oysters, and clams.

The largest marine fishery in the world is the anchoveta or Peruvian anchovy (*Engraulis ringens*) fishery, which can account for about 8 per cent or more of the global marine catch of seafood in any particular year. Anchoveta are small (up to about 20 cm in length) fast-growing fish that filter-feed on phytoplankton and zooplankton in the nutrient-rich upwelling areas off the coast of Peru. They tend to congregate in large, dense schools, which allows them to be caught efficiently in large numbers by purse seiners and pelagic trawlers.

The anchoveta catch fluctuates wildly from year to year. In a good year catches can be upwards of 11 million tonnes; in bad years the catch can be around 2 million tonnes; and in very bad years below 150,000 tonnes. Much of this fluctuation is related to the El Niño Southern Oscillation (ENSO) (see Chapter 2).

The second largest fishery in the world is for Alaska pollock which is a demersal species widely distributed in the North Pacific Ocean. It is caught using pelagic trawls which occasionally contact the bottom but with minimal damage compared to bottom trawls. Catches are currently averaging about 3 million tonnes annually. This species is widely used in the fast food industry. During El Niño years the catch of Alaska pollock can be greater than for anchoveta.

The third largest fishery is for skipjack tuna, which are abundant in the tropical regions of the Atlantic, Pacific, and Indian oceans and are the main species of canned tuna. They are often caught by purse seiners and catches have been averaging about 2.6 million tonnes annually.

The anchoveta and some other very oily fish are less desirable for direct consumption by humans, but their high oil content makes them ideal to produce fish meal and fish oil, which are both valuable commodities. Fish meal is made by cooking, drying, and grinding up the fish and is used as a protein supplement incorporated into commercial feeds for farmed fish, poultry, and pigs. Fish oil is pressed from the cooked fish and is used mainly in the production of feed for farmed fish, but is also put into capsules as a human health supplement. Almost the entire Peruvian anchoveta catch is converted into fish meal and fish oil, which accounts for about a third of the global production of these products.

The inclusion of wild-caught fish products into feeds used to farm seafood has been controversial because analyses have shown there is a considerable loss of food energy compared to the fish being eaten directly by humans. However, progress is being made in addressing this issue and the amount of fish being converted into fish meal and oil has been declining. In the 1990s about a third of the entire global catch of fish was converted into these ingredients rather than consumed directly by humans, but in 2016 about 15 million tonnes, or about 19 per cent of the reported marine catch, was converted into fish meal and oil. Much more of these ingredients are now being produced from fish by-products such as offal and fish frames, which were often wasted in the past, and they are being used more selectively in fish feeds. Some recent work has shown that under current marine fish farming practices 1 kilogram of wild fish ingredients incorporated into fish feed is producing on average about 2 kilograms of farmed fish, in which case the use of these wild fish ingredients is producing additional food for humans. For salmon farming, specifically, this ratio is currently less favourable, about 1:1, but improving over time with the adoption of improved practices.

## Trends in the global seafood catch—exceeding the limits of exploitation

The Food and Agriculture Organization of the United Nations (FAO) maintains a global database on seafood catches going back to 1950 which is based on official information reported by member countries and the European Union. FAO data show that in 1950 the total global catch of marine seafood was 18 million metric tonnes fresh weight. Catches increased steadily and rapidly from then until the late 1980s when they began to level off (see Figure 39). The highest reported global catch on record was

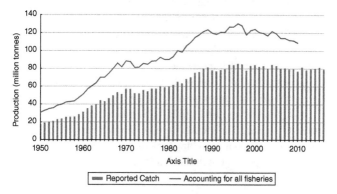

**39. Trends in annual global marine capture fisheries production. Bars show reported catch while line shows estimated catch accounting for all fisheries.**

86 million tonnes in 1996. Since then global marine catches have fluctuated below this level and the reported catch in 2016 was 79 million tonnes. But these data do not include all marine seafood caught globally. This is because official member reports often omit bycatch that is frequently discarded at sea, as well as catches from illegal, unreported, and unregulated (IUU) fisheries, including small-scale artisanal and subsistence fisheries, all of which are difficult to obtain reliable information for. Thus, the actual catch of marine species from the wild will be higher than the FAO estimates, but by how much is controversial. Recent studies drawing from a broader range of data sources suggest that the global catch most likely peaked at around 130 million tonnes in the 1990s and that catches have been declining since then to somewhere between 100 and 120 million tonnes currently.

Looking ahead, what contribution will the Global Ocean make to providing wild-caught seafood for the roughly 9.8 billion people that will occupy our planet by 2050? One way to address this question is founded on basic ecological principles regarding the efficiency with which the Global Ocean's total net primary production is transferred through marine food chains (see Table 2). Using this approach, the Global Ocean can be divided into three provinces designated by their different levels of primary productivity. The largest, the oceanic province, consists of all the open-ocean regions of the Global Ocean. It has the lowest average primary productivity but occupies the greatest area by far—about 90 per cent of the total area of the Global Ocean. The second is the coastal province encompassing all the coastal marine waters. It has higher primary productivity but a much smaller area—just under 10 per cent. The third is the upwelling province that includes all the major upwelling regions. It occupies a tiny area—about 1 per cent of the Global Ocean—but has very high average primary productivity.

The next step is to estimate the number of trophic levels within the food webs in each province. It is reasonable to assign an

**Table 2.** Estimated annual total fish production of the Global Ocean (from Ryther 1969)

| Province | % of the ocean | Mean productivity ($g C m^{-2} yr^{-1}$) | Total primary production (billion tonnes of $C yr^{-1}$) | Number of trophic levels | % efficiency | Fish production (tonnes fresh weight) |
|---|---|---|---|---|---|---|
| Oceanic | 90.0 | 50 | 16.3 | 5 | 10 | 1.6 million |
| Coastal | 9.9 | 100 | 3.6 | 3 | 15 | 120 million |
| Upwelling | 0.1 | 300 | 0.1 | 1.5 | 20 | 120 million |
| Total | | | | | | Approximately 240 million |

average of five trophic levels between the primary producers and exploited fish in complex open ocean food webs (see Chapter 2). By comparison, food webs in upwelling regions are very simple. The anchoveta fishery is a case in point—here the anchoveta feed directly on phytoplankton or on zooplankton, so they are only one or two trophic levels removed from the primary producers. Thus, this province can be assigned 1.5 trophic levels between primary producers and fish. The coastal province can be viewed as transitional in nature between the open ocean and the upwelling regions and can be assigned an average of three trophic levels.

One can then estimate the efficiency with which energy is transferred between each trophic level in each province. Energy transfer is least efficient in the complex and dispersed food webs of the open ocean and can be assigned a value of 10 per cent. In contrast, energy transfer is much more efficient in the simple, concentrated food webs of the upwelling zones and can be assigned a figure of 20 per cent. An intermediate figure of 15 per cent can be assigned to the coastal province.

By combining the data on primary productivity, number of trophic levels, and energy transfer efficiencies in each province, it is possible to estimate the approximate annual fish production in each province and thus produce an estimate of total world fish production—240 million metric tonnes per year (see Table 2).

One of the insights from Ryther's analysis is the overwhelming importance of coastal and upwelling regions in fish production because of their high primary productivity and simpler and more efficient food webs. Together they account for almost all the fish production of the Global Ocean. The open-ocean region, despite its immense size, accounts for little fish production. This is consistent with the known facts—the largest commercial fisheries—for clupeoids, gadoids, and flatfish—are all situated in upwelling regions and coastal areas of the Global Ocean.

The exception is the open-ocean fisheries for species such as tuna and mackerel, the top predators in a vast oceanic system.

Not all the estimated 240 million tonnes of annual fish production is available to humans for harvest. If all of it were exploited each year fish stocks would be rapidly depleted to unsustainable levels. Also, much of this fish production must be shared with other top predators besides humans—such as marine mammals, sharks, and seabirds—and it would be impossible to locate and exploit all fish stocks at optimal levels at all times. On this basis, it is reasonable to assume that a maximum of about 100 million tonnes of fish would be available for exploitation by humans on a sustained basis each year.

These and other similar approaches provide a theoretical underpinning for estimating the amount of seafood that humans might expect to sustainably capture from the Global Ocean based on ecological principles. The key point is that the actual global seafood catch, which is currently estimated at about 120 million tonnes, exceeds the theoretical limits. Furthermore, global catches are now stagnant or in gradual decline despite an overall increase in fishing effort in terms of the size and efficiency of the fishing fleet. These are clear signals that globally marine fisheries are now overexploited. There is thus little, if any, headroom for increasing the amount of wild-caught seafood humans can extract from the oceans to feed a burgeoning human population under our current fisheries management approaches.

This conclusion is further supported by the increasingly precarious state of global marine fishery resources. The most recent information from the FAO shows that 60 per cent of all fish stocks are fully exploited—their current catches are at their maximum sustainable levels of production and there is no scope for further expansion. Another 33 per cent are overfished and in decline or facing collapse (up from 10 per cent in 1974), and only 7 per cent are underexploited. This paints a grim picture of the state of the world's marine fisheries and the real question is not whether we can catch more fish from the oceans but whether we can responsibly sustain the amount of fish we are harvesting at present, which represents a vital source of food for the human population. Clearly, the 70 per cent or so extra food soon needed to feed around 9.8 billion people will have to come from somewhere else. Marine aquaculture is currently helping to fill the demand for seafood—in 2016 a total of 29 million tonnes of farmed seafood was produced in addition to the roughly 120 million tonnes of wild-caught seafood.

Overfishing not only devastates target species, but also impacts the functioning of the marine ecosystems they are part of, often in unexpected ways. The collapse of the cod fishery in the north-west Atlantic provides a good example of how the effects of overfishing can ripple down dramatically through the rest of the food web. Cod were once abundant off the north-eastern coast of the USA but the fishery collapsed due to overfishing in the early 1990s. Cod are a major predator of lobsters and with their removal from the ecosystem lobster numbers exploded. There is so far no sign of the cod fishery recovering, so this transition to a lobster-dominated ecosystem appears to be a new long-term state. The high numbers of lobsters are good for lobster fishers at this time but crowding, together with stress associated with ocean warming, makes them susceptible to disease, which has decimated lobster stocks in other areas. Another example of the ecosystem effects of overfishing involves the sardine and anchovy fisheries off the south-west coast of Africa. These fisheries collapsed in the 1970s as a result of a combination of overfishing and changing environmental conditions. The sardines and anchovies were rapidly replaced with large numbers of goby fish and jellyfish, which are much less nutritious. As a result, other animals in the ecosystem, which formerly fed on nutritious sardines and anchovies, declined precipitously, including penguins and gannets, as well as several species of hake that were once thriving fisheries.

This also appears to be a long-term transition to a new and less diverse and productive ecosystem. These examples show that fisheries managers must not only manage the target species but also take a much broader 'ecosystems-based' approach to fisheries management which requires a much greater understanding of how all the organisms at various trophic levels in the ecosystem interact.

## Future of marine fisheries—the last wild food

It is currently proving very difficult for nations to manage overexploitation and control illegal fishing in the face of the ever-increasing demand for valuable wild-caught seafood fuelled by a rapidly growing and affluent global population. Governments around the world must work much harder to better monitor stock levels, enforce catch limits, reduce fishing effort, and stamp out illegal fishing to allow stocks to rebuild and ecosystems to hopefully recover. They also must be prepared to make hard and unpopular decisions to close fisheries that are not managed in a sustainable fashion. The retailers and consumers of seafood have an important role to play as well. Overexploitation of fish stocks is often driven by consumer demand, indifference, and lack of knowledge. If consumers are properly educated about the problems of overfishing, they will be in a better position to make decisions not to purchase unsustainably caught species. For example, the Marine Stewardship Council (MSC) certifies fisheries deemed to be sustainably fished by independent assessment by marine scientists. Packaged seafood products from MSC certified fisheries bear an easily recognizable MSC 'blue fish' label. Similarly, WWF produces country-specific guides that provide information on the state of individual fish stocks and labels species as green, yellow, or red depending on their level of sustainability.

There has been some recent progress in rebuilding overfished stocks, particularly in developed countries with improved management regimes. For example, the proportion of stocks fished sustainably in US waters increased from 53 per cent in 2005 to 74 per cent in 2016 and in Australian waters from 27 per cent in 2004 to 69 per cent in 2015. But overfishing is increasing in less developed countries due to limited monitoring, management, and enforcement capacity. Reversing this trend and rebuilding depleted stocks is a hugely important enterprise for maintaining the diversity and health of marine ecosystems and for preserving humankind's last significant source of wild food.

## The role of marine protected areas in fisheries management

Properly enforced 'no-take' marine protected areas (MPAs) provide a space within which fish and invertebrates cannot be harvested by recreational and commercial fishers. This helps species build up their numbers within the MPA, live longer, and reach a larger body size. Larger animals can produce a greater number of eggs, while the longer an animal lives, the more times it can reproduce. Thus, dense populations of large-bodied, long-lived animals in MPAs produce large numbers of eggs and larvae that drift out of the reserve and enhance recruitment to depleted stocks in adjacent areas. There is also a 'spillover' effect created by MPAs. As populations of mobile species, such as fish and lobsters, grow in number they eventually saturate the protected area and begin to spread out into unprotected areas where they bolster recreational and commercial fisheries. It is not surprising, then, as many fishers have come to realize, that the best fishing is often adjacent to an established MPA.

A key question is how much of the Global Ocean needs to be protected to bring about a meaningful recovery of our planet's ocean ecosystems and permit fish stocks to rebuild to the point where they provide stable commercial returns to the fishing industry and a sustainable and secure supply of wild-caught seafood for humans. The growing consensus among marine scientists is that a network of no-take MPAs covering more than 30 per cent of the area of the Global Ocean needs to be established and effectively managed to achieve this, although some ecologists now believe that half of the Earth's land and, by extension, half of the oceans, needs to be protected from human interference to stem the loss of our planet's biodiversity.

The costs of establishing and administering such a network of MPAs are high but would be more than offset by the expected benefits, including an uplift in yield of commercial fisheries, the large numbers of jobs created, and the increased income from visitors to MPAs. For instance, fisheries scientists have estimated that if all overfished marine stocks were allowed to rebuild and then fished sustainably, the global catch could increase by about 16.5 million tonnes annually.

In 2015 the United Nations adopted a Sustainable Development Goal for the Oceans to 'Conserve and sustainably use the oceans, seas and marine resources'. One of the targets of the goal is that at least 10 per cent of coastal and marine areas should be protected by 2020. An online platform, Protected Planet, managed by the UN's World Conservation Monitoring Centre, provides monthly updated statistics on the coverage of MPAs globally. Their data show that MPA coverage in 2000 was only 2 million square kilometres, or 0.7 per cent of the oceans. Since then there has been over a tenfold increase with over 15,000 MPAs established worldwide covering 27 million square kilometres or about 7.5 per cent of the oceans. This is good progress, but still well short of the minimum 30 per cent required. Furthermore, the coverage is uneven, with less than 3 million square kilometres of coverage, or about 11 per cent of the total, in the High Seas. This is because of the legal difficulties currently involved in creating MPAs in international waters. Another problem is that due to a lack of resources many MPAs are not effectively managed so fishing carries on as before. In response, the International Union for Conservation of Nature is setting up a 'Green List' of MPAs that meet the required management standards to deliver real conservation outcomes.

MPAs will not address all the problems of the oceans, of course, and are just one of the tools for their effective management. Apart from overfishing, marine ecosystems are under pressure from pollution and the climate change effects of ocean warming and acidification, all of which ignore the boundaries of marine reserves and work synergistically to reduce the ability of marine ecosystems to function normally. Other responses and tools are needed to address these issues.

# Chapter 9
# The future of our oceans

As discussed throughout this book, Anthropocene marine ecosystems are profoundly altered compared to their state in the pre-industrial era and the pace of change has accelerated greatly over the past four decades. What will be the state of our oceans in 2050? Are we destined to stay on the current negative course? Or are we capable of plotting a new course and reversing many of the most severe impacts we have had on the oceans so far?

## The end of ocean wilderness

A recent study of human impacts on global ocean ecosystems has revealed the full extent of human influence on the marine environment and acts as a 2018 baseline against which to measure future changes, be they positive or negative. This analysis involved identifying areas of 'marine wilderness'—places where marine ecosystems are functioning in a largely natural way mostly free from human disturbance. The researchers divided the Global Ocean into fifteen geographic regions and scored each region based on the number, intensity, and cumulative effects of a range of human 'stressors' including pollution, overfishing, invasive species, and climate change. They discovered that when climate change stressors (which included rising ocean temperatures and acidification) were included in the analysis no ocean wilderness remained anywhere on the planet since the impacts of climate change are so widespread and unmanageable. Even when climate change stressors were not included, only about 13 per cent of the oceans could be considered as marine wilderness. These areas were mainly located in parts of the open ocean in the southern hemisphere and in the Southern Ocean and Arctic Ocean. The painful conclusion from this and other studies is that human activities have impacted the oceans everywhere and the notion that there are still marine ecosystems that are intact and can be 'conserved' is no longer tenable. The challenge for humans now is to find ways to stabilize and prevent the further degradation of marine ecosystems and embark on a path of ocean restoration. It remains to be seen whether society will fully understand and acknowledge the problems facing the oceans and care enough about the marine environment to put in place quickly enough the measures that are required to restore marine ecosystems. Not to do so, however, will ultimately be calamitous. The actions humans take in the next few decades will decide the outcome.

I am optimistic that society and its institutions will respond rationally and choose a path that puts us on a course to restore our oceans and protect the many benefits they provide. This position is based on the knowledge that public and media awareness, concern, and engagement in marine environmental matters are increasing rapidly, particularly among young people, and also spreading to less developed countries. These well-informed people will put increased pressure on politicians, international and national organizations, and the companies they do business with to deal responsibly with marine environmental issues. Furthermore, some of these people will become political leaders and prominent spokespeople for the marine environment, while others will be part of a new wave of marine scientists who can set us on a path to restore marine ecosystems. Keeping this in mind, I will briefly lay out a vision for the state of the Global Ocean in 2050 in ten points below. It is a mainly hopeful outlook but one that is entirely realistic if we accelerate and fully implement many of the actions already under way and are bold with implementing some new approaches.

# The oceans in 2050—one possible future

In 2050 the majority of the world's nations are active members of an **effective international body** responsible for the overall governance of Earth's oceans. This organization has evolved from the International Union for Conservation of Nature and operates under an international legal framework first provided by the United Nations Convention on the Law of the Sea (UNCLOS). It develops and implements long-term strategies for the management of the planet's marine environment and its resources, and coordinates and monitors the actions and progress of its members in implementing these strategies. It also provides effective enforcement in the case of non-compliance. Members of this body have set aside most of their disagreements on ocean management issues that were based on short-term priorities and national self-interest to cooperate effectively to achieve agreed long-term outcomes for ocean ecosystem restoration and resource management.

The **marine plastic debris crisis** has been largely addressed thanks to a groundswell of community concern that galvanized governments and industry to take appropriate measures. Governments, particularly in poorer nations, have funded improved waste collection systems that stem the loss of plastics from the land to the oceans via rivers. The demand for plastics has been greatly reduced because of global bans on single use plastic bags and reduced use of plastic in packaging. Furthermore, efficient systems are in place to collect and recycle plastic into new products thus greatly reducing the amount of virgin plastic being manufactured. After some initial teething problems effective systems to remove plastic debris from the oceanic gyres are in operation and the amount of plastic in the gyres has been reduced by three-quarters, reducing harm to marine organisms and reducing the source of much of the microplastics in the oceans. Shorelines and seascapes around the world are now much cleaner and there are much reduced levels of microplastics in seafood.

**Nutrient losses** to the oceans have been halved as a result of widespread adoption of more sustainable agricultural practices in most countries driven by consumer concerns, effective legislation, and improved education and decision support tools for farmers. In addition, most coastal cities have installed effective sewage management systems. These initiatives have greatly reduced the loss of nitrogen and phosphorus into the coastal environment and the number and size of dead zones in the coastal oceans are decreasing for the first time. The frequency of harmful phytoplankton blooms is also decreasing.

A global network of actively managed and properly enforced **no-take marine protected areas** (MPAs) is in place covering more than 30 per cent of the area of the Global Ocean, with half of this coverage in the High Seas now that many of the legal difficulties involved in creating MPAs in international waters have been resolved. The cost of administering this network is greatly offset by a significant uplift in yield and income from commercial fisheries benefiting from the fisheries enhancement effects provided by MPAs. As a result, many large fishing companies sponsor the ongoing management of MPAs. MPAs receive significant income from marine tourists snorkelling and scuba diving in their protected waters. The MPA network has become so successful and widely supported that the international ocean governance body has set a new target of 50 per cent Global Ocean coverage by 2070.

**Improved fisheries management practices** based on an ecosystems approach are the norm for most countries, with wealthy countries assisting less wealthy countries to develop, implement, and enforce improved fisheries management programmes and apply them to previously unregulated fisheries. There is much improved monitoring of stock levels and enforcement of catch limits among all marine fishing nations. Country reports on marine seafood catches to the FAO now include data on bycatch as well as estimated catches from artisanal and recreational fisheries, making the FAO annual reports more representative of actual global marine catches. Illegal fishing has been almost completely abolished as a result of detection and tracking of all fishing boats on the oceans using satellite radar data obtained from a large fleet of CubeSats orbiting the planet. Global fishing effort has been substantially reduced with people formerly working in the fishing industry employed in monitoring and management of MPAs and marine tourism. The fishing industry has moved to a 'take less and earn more' business model with the emphasis on obtaining better prices for quality fish products harvested sustainably. Bottom trawl fishing has been reduced to a few small approved and monitored areas of ocean floor. All large food retailers supply only seafood products sourced from fisheries certified as sustainable by approved organizations. Well-informed consumers routinely use 'smart codes' on packaged seafood to confirm its origin. For the first time in many decades many formerly depleted fish stocks are rebuilding and the biodiversity and stability of a range of marine ecosystems are improving as the numbers of ecologically important marine animals such as predatory fish, whales, and sharks recover.

The Global Ocean is **monitored intensively** using a vast array of sensors mounted on ocean moorings, robotic underwater gliders, autonomous underwater vehicles, and satellites. These provide huge amounts of real time data on the physical, chemical, and biological status of the marine environment that are analysed and visualized using advanced computing techniques and made widely available as free and open data sources. This information provides a detailed synoptic picture of the status of the oceans updated monthly which informs ongoing management decisions and allows measurement of progress towards agreed ocean restoration goals.

Much work is now directed at **adapting to and managing the ongoing impacts of the human-induced climate crisis** on marine ecosystems and fisheries. Although many countries have made progress in decarbonizing their economies, the world is struggling to meet the emission reduction goals of the 2015 Paris Agreement to keep the increase in average global air temperature to well below 2°C. The planet is on course for a 2.5°C warming before the climate is stabilized sometime after 2100 and greenhouse gases in the atmosphere start to decline. As predicted, more than 80 per cent of the planet's coral reefs have collapsed into macroalgal-dominated systems as a result of a series of global back-to-back bleaching events caused by ocean surface temperature extremes. The remaining patches of relatively healthy coral reefs surviving in climate change refugia in scattered places in the tropical oceans are carefully managed and protected from overfishing, pollution, and crown-of-thorns infestations so that they remain as resilient as possible to ongoing ocean temperature rises, ocean acidification, and the increased frequency and intensity of storms. It is hoped that the surviving corals at these sites will repopulate neighbouring regions once the planet's climate has stabilized. Active interventions are well under way to partially restore coral reef ecosystems at some key sites. These include assisted gene flow and assisted evolution technologies. Gene editing techniques are being used to engineer strains of corals that have greater tolerance to thermal stress and acidified seawater. These are cultivated in large numbers in land-based facilities and outplanted to favourable sites to partially resurrect some coral reef ecosystems and restore some of their ecosystem services. Engineered artificial reef-like habitats are being constructed in many places to re-establish some coral reef fisheries and help protect vulnerable coastlines.

**Offshore oil and gas exploration has been greatly reduced** as these fossil fuels have been substantially replaced by renewable energy sources deployed on land and in the oceans. This has greatly reduced marine pollution from exploration and transport of oil. The electrification of the transport system has also greatly reduced input of oil into the coastal environment from leakage associated with the consumption of oil-derived products in cars and trucks.

The **industrialization of the oceans** for seabed mining, offshore wind and wave energy developments, and ocean farming is accelerating, but these activities are assessed and managed carefully. Exploration of deep-ocean ecosystems is well advanced using autonomous underwater vehicles and a new generation of crewed deep-ocean vehicles. This knowledge is used to assess the impacts of seabed mining proposals. Climate engineering projects involving large-scale iron fertilization of the oceans have been abandoned as too ecologically risky.

Governments and industry invest heavily in **marine research and development** to provide the knowledge needed to adapt to climate change disruptions of marine ecosystems and restore the marine environment and its resources. Marine biologists, ecologists, molecular biologists, microbiologists, chemists, physical oceanographers, sensor and robotics engineers, computer scientists, and information system experts work together on complex marine environmental issues.

In conclusion, this scenario for the future of the oceans in 2050 depicts just one possible pathway. Many other scenarios are possible, of course, including some that are not nearly so hopeful. The next decade will be the one in which society decides the state of life in the oceans that we leave behind for future generations.

# References and further reading

## General reading

Knowlton, N. (2010) *Citizens of the Sea: Wondrous Creatures from the Census of Marine Life*. National Geographic Society.

Levinton, J. S. (2018) *Marine Biology: Function, Biodiversity, Ecology*. 5th Edition. New York: Oxford University Press.

Palumbi, S. R. and Palumbi, A. R. (2014) *The Extreme Life of the Sea*. Princeton and Oxford: Princeton University Press.

Roberts, C. (2012) *Ocean of Life*. London: Penguin Group.

Rossi, S. (2019) *Oceans in Decline*. Basel: Springer Nature Switzerland AG.

Thomas, D. N. and Bowers, D. G. (2012, reprinted 2018) *Introducing Oceanography*. Edinburgh: Dunedin Academic Press Ltd.

## Introduction

Barbier, E. B. (2017) Marine ecosystem services. *Current Biology* 27, R507–10.

Hoegh-Guldberg, O. et al. (2015) *Reviving the Ocean Economy: The Case for Action—2015*. Geneva: WWF International.

United Nations Department of Economic and Social Affairs (2017) World population projected to reach 9.8 billion in 2050, and 11.2 billion in 2100. (Online) Available from: <https://www.un.org/development/desa/en/news/population/world-population-prospects-2017.html> (Accessed 11 January 2019).

WWF-UK [GB]. (2018) Introducing the Sustainable Blue Economy
Finance Principles. (Online) Available from: <https://www.wwf.
org.uk/updates/sustainable-blue-economy-finance-principles>
(Accessed 11 January 2019).

## Chapter 1: The oceanic environment

Bähr U. (ed.) (2017) *Ocean Atlas: Facts and Figures on the Threats to
our Marine Ecosystems.* 1st Edition. Berlin: Heinrich Boll
Foundation.

Charette, M. A. and Smith, W. H. F. (2010) The volume of the Earth's
ocean. *Oceanography* 23 (2), 104–6.

Laffoley, D. and Baxter, J. M. (eds). (2016) *Explaining Ocean
Warming: Causes, Scale, Effects and Consequences. Full Report.*
Gland, Switzerland: IUCN.

Praetorius, S. K. (2018) North Atlantic circulation slows down. *Nature*
556, 180–1.

Quote Investigator (2017) Planet 'Earth': we should have called it 'Sea'.
(Online) Available from: <https://quoteinvestigator.
com/2017/01/25/water-planet/> (Accessed 11 January 2019).

USGS (2016) How much water is there on, in, and above the Earth?
(Online) Available from: <https://water.usgs.gov/edu/
earthhowmuch.html> (Accessed 11 January 2019).

Wessel, P., Sandwell, D. T., and Kim, S.-S. (2010) The global seamount
census. *Oceanography* 23 (1), 24–33.

Wittmann, A. C. and Pörtner, H.-O. (2013) Sensitivities of extant animal
taxa to ocean acidification. *Nature Climate Change* 3, 995–1001.

## Chapter 2: Marine biological processes

Bar-On, Y. M., Phillips, R., and Milo, R. (2018) The biomass distribution
on Earth, *Proceedings of the National Academy of Sciences* (Online)
115 (25) 6506–11. Available from: DOI:10.1073/pnas.1711842115
(Accessed 12 January 2019).

Biller, S. J., Berube, P. M., Lindell, D., and Chisholm, S. W. (2015)
*Prochlorococcus*: the structure and function of collective diversity.
*Nat Rev Microbiol.*13 (1), 13–27. Available from: DOI: 10.1038/
nrmicro3378.

Brierley, A. S. (2017) Plankton. *Current Biology* 27, R478–83.

Bristow, L. A., Mohr, W., Ahmerkamp, S., and Kuypers, M. M. M. (2017) Nutrients that limit growth in the oceans. *Current Biology* 27, R474–8.

Carradec, Q., Pelletier, E., Da Silva, C., et al. (2018) A global ocean atlas of eukaryotic genes. *Nature Communications* (Online) 9, Article number 373. Available from: <https://doi.org/10.1038/s41467-017-02342-1> (Accessed 19 January 2019).

Chisholm, S. W. (2017) *Prochlorococcus. Current Biology* 27, R447–8.

Coutinho, F. H., Silveira, C. B., Gregoracci, G. B., et al. (2017) Marine viruses discovered via metagenomics shed light on viral strategies throughout the oceans. *Nature Communications* (Online) 8, 15955. Available from: doi: 10.1038/ncomms15955 (Accessed 19 January 2019).

de Vargas, C., Audic, S., Henry, N., et al. (2017) Eukaryotic plankton diversity in the sunlit ocean. *Science* (Online) 348 (6237). Available from: DOI: 10.1126/science.1261605. (Accessed 19 January 2019).

Fenchel, T. (2008) The microbial loop—25 years later. *Journal of Experimental Marine Biology and Ecology* 366, 99–103.

Keeling, P. J. and del Campo, J. (2017) Marine protists are not just big bacteria. *Current Biology* 27, R541–9.

NASA (n.d.). *Oceancolor Web.* (Online) Available from: <https://oceancolor.gsfc.nasa.gov/about/> (Accessed 13 January 2019).

Pomeroy, L. R. (1974) The ocean's food web, a changing paradigm. *BioScience* 24 (9), 499–504.

Pomeroy, L. R., Williams, P. J. leB., Farooq Azam, F., and Hobbie, J. E. (2007) The microbial loop. *Oceanography* 20 (2), 28–33.

Powell, H. (2008) Fertilizing the ocean with iron. *Oceanus Magazine* 46 (1), 4–9. Available from: <https://www.whoi.edu/oceanus/feature/fertilizing-the-ocean-with-iron>.

Salazar, G. and Sunagawa, S. (2017) Marine microbial diversity. *Current Biology* 27, R489–94.

Suttle, C. A. (2007) Marine viruses—major players in the global system. *Nature Reviews Microbiology* 5, 801–12.

Tara Oceans (2015) Planktonic world: the new frontier. First scientific results from the Tara Oceans expedition. (Online) Available from: <https://oceans.taraexpeditions.org/wp-content/uploads/2015/05/press-kit_tara-oceans.pdf> (Accessed 11 January 2019).

Tollefson, J. (2017) Plankton-boosting project in Chile sparks controversy. *Nature* 345, 393–4.

## Chapter 3: Life in the coastal ocean

Anderson, D. (2014) HABs in a changing world: a perspective on harmful algal blooms, their impacts, and research and management in a dynamic era of climactic and environmental change. *Harmful Algae 2012 (2012)*, 3–17. Available from: <https://www.ncbi.nlm.nih.gov/pmc/articles/PMC4667985/>.

Breitburg, D., Grégoire, M., and Isensee, K. (eds). (2018) *Global Ocean Oxygen Network 2018. The Ocean is Losing its Breath: Declining Oxygen in the World's Ocean and Coastal Waters.* IOC-UNESCO, IOC Technical Series, No. 137 40pp. Available from: <http://www.fao.org/fishery/topic/14776/en>.

De Poorter, M., Darby, C., and MacKay, J. (2009) *Marine Menace: Alien Invasive Species in the Marine Environment.* Gland, Switzerland: IUCN, 1–31.

Eriksen, M., Lebreton, L. C. M., Carson, H. S., Thiel, M., Moore, C. J., et al. (2014) Plastic pollution in the world's oceans: more than 5 trillion plastic pieces weighing over 250,000 tons afloat at sea. *PLOS ONE* (Online) 9 (12). e111913. Available from: <https://doi.org/10.1371/journal.pone.0111913> (Accessed 13 January 2019).

Evans, S. M., Griffin, K. J., Blick, R. A. J., Poore, A. G. B., and Vergés A. (2018) Seagrass on the brink: decline of threatened seagrass *Posidonia australis* continues following protection. *PLoS ONE* (Online) 13 (4): e0190370. Available from: <https://doi.org/10.1371/journal.pone.0190370> (Accessed 13 January 2019).

Food and Agriculture Organization of the United Nations (2019) Ballast waters: pollution and invasive species. (Online) Available from: <http://www.fao.org/fishery/topic/14776/en> (Accessed 13 January 2019).

Galloway, T. and Lewis, C. (2017) Marine microplastics. *Current Biology* 27, R445–6.

Irwin, A. (2018) How to solve a problem like plastics. *New Scientist* 238 (3178), 25–31.

Ling, S. D., Scheibling, R. E., Rassweiler, A., et al. (2015) Global regime shift dynamics of catastrophic sea urchin overgrazing. *Phil. Trans. R. Soc. B: Biological Sciences* (Online). Available from: <http://dx.doi.org/10.1098/rstb.2013.0269> (Accessed 13 January 2019).

McClenachan, L., Jackson, J. B. C., and Newman, M. J. H. (2006) Conservation implications of historic sea turtle nesting beach loss. *Front. Ecol. Environ.* 4 (6), 290–6.

The Ocean Cleanup (2019) The largest cleanup in history. (Online)
Available from: <https://www.theoceancleanup.com/> (Accessed
13 January 2019).

Reynolds, P. L. (2018) Seagrass and seagrass beds. (Online)
Available from: <https://ocean.si.edu/ocean-life/plants-algae/
seagrass-and-seagrass-beds> (Accessed 13 January 2019).

Tegner, M. J. and Dayton, P. K. (2000) Ecosystem effects of fishing in
kelp forest communities. *ICES Journal of Marine Science* 57,
579–89.

Unsworth, R. K. F. and Cullen-Unsworth, L. C. (2017) Seagrass
meadows. *Current Biology* 27, R443–5.

## Chapter 4: Polar marine biology

Alfred-Wegener-Institut (2018) What goes on beneath the floes.
(Online) Available from: <https://www.awi.de/en/focus/sea-ice/
life-in-and-underneath-sea-ice.html> (Accessed 13 January 2019).

Atkinson, A., Siegel, V., Pakhomov, E., and Rothery, P. (2004)
Long-term decline in krill stock and increase in salps within the
Southern Ocean. *Nature* 432, 100–3.

Barnes, D. K. A. and Tarling, G. A. (2017) Polar oceans in a changing
climate. *Current Biology* 27, R454–60.

Bowman, J. S. (2015). The relationship between sea ice bacterial
community structure and biogeochemistry: a synthesis of current
knowledge and known unknowns. *Elementa: Science of the
Anthropocene* (Online) 3, 000072. Available from: DOI:http://doi.
org/10.12952/journal.elementa.000072 (Accessed 14 January
2019).

FAO (2018) *The State of World Fisheries and Aquaculture 2018—
Meeting the Sustainable Development Goals.* Rome. Licence:
CC BY-NC-SA 3.0 IGO.

Kock, K.-H. (2007) Antarctic marine living resources—exploitation
and its management in the Southern Ocean. *Antarctic Science*
19 (2), 231–8.

Miller, D. G. M. (1991) Exploitation of Antarctic marine living
resources: a brief history and a possible approach to managing the
krill fishery. *South African Journal of Marine Science* 10, 321–39.

National Snow & Ice Data Center (2018) State of the cryosphere: is
the cryosphere sending signals about climate change? (Online)
Available from: <https://nsidc.org/cryosphere/sotc/sea_ice.html>
(Accessed 13 January 2019).

Anthony, K., Bay, L. K., Costanza, R., Firn, J., et al. (2017) New interventions are needed to save coral reefs. *Nature Ecology & Evolution* 1, 1420–2.

Beyer, H. L., Kennedy, E. V., Beger, M., et al. (2018) Risk-sensitive planning for conserving coral reefs under rapid climate change. *Conservation Letters* (Online) 11 (6) e12587. Available from: <https://doi.org/10.1111/conl.12587> (Accessed 14 January 2019).

Camp, E. F., Schoepf, V., Mumby, P. J., Hardtke, L. A., et al. (2018) The future of coral reefs subject to rapid climate change: lessons from natural extreme environments. *Front. Mar. Sci.* (Online) 5 (4). Available from: doi: 10.3389/fmars.2018.00004 (Accessed 14 January 2019).

Degan, B. (2017) Love connection: breakthrough fights crown-of-thorns starfish with pheromones. (Online) Available from: <https://theconversation.com/love-connection-breakthrough-fights-crown-of-thorns-starfish-with-pheromones-75779> (Accessed 14 January 2019).

Guest, J. R., Edmunds, P. J., Gates, R. D., et al. (2018) A framework for identifying and characterising coral reef 'oases' against a backdrop of degradation. *J Appl Ecol.* 55 (6), 2865–75. Available from: <https://doi.org/10.1111/1365%962664.13179>.

Hall, M. R., Kocot, K. M., Baughman, K. W., Fernandez-Valverde, S. L., et al. (2017) The crown-of-thorns starfish genome as a guide for biocontrol of this coral reef pest. *Nature* 544 (7649), 231–4. Available from: doi: 10.1038/nature22033.

Hausheer, J. E. (2018) River pollution threatens Australia's great barrier reef. (Online) Available from: <https://blog.nature.org/science/2018/12/03/river-pollution-threatens-australias-great-barrier-reef/> (Accessed 19 January 2019).

Hughes, T. P., Barnes, M. L., Bellwood, D. R., Cinner, J. E., et al. (2017) Coral reefs in the Anthropocene. *Nature* 546, 82–90.

Keith, S. A., Maynard, J. A., Edwards, A. J., et al. (2016) Coral mass spawning predicted by rapid seasonal rise in ocean temperature. *Proc. R. Soc. B* (Online) 283: 20160011. Available from: <http://dx.doi.org/10.1098/rspb.2016.0011> (Accessed 19 January 2019).

Libro, S. and Vollmer, S. V. (2016) Genetic signature of resistance to White Band Disease in the Caribbean Staghorn Coral *Acropora cervicornis*. *PLoS ONE* (Online)11 (1): e0146636. Available from: doi:10.1371/journal. pone.0146636 (Accessed 14 January 2019).

Mollica, N. R., Guo, W., Cohen, A. L., Huang, K.-F., et al. (2018) Ocean acidification affects coral growth by reducing skeletal density. *PNAS* (Online)115 (8), 1754–9. Available from: <http://www.pnas.org/cgi/doi/10.1073/pnas.1712806115> (Accessed 14 January 2019).

Plaisance, L., Caley, M. J., Brainard, R. E., and Knowlton, N. (2011) The diversity of coral reefs: what are we missing? *PLoS ONE* (Online) 6 (10): e25026. Available from: doi:10.1371/journal.pone.0025026 (Accessed 14 January 2019).

Pratchett, M. S., Caballes, C. F., Wilmes, J. C., Matthews, S., et al. (2017). Thirty years of research on Crown-of-Thorns Starfish (1986–2016): scientific advances and emerging opportunities. *Diversity* (Online) 9 (4), 41. Available from: doi:10.3390/d9040041 (Accessed 14 January 2019).

Putman, H. M., Barott, K. L., Ainsworth, T. D., and Gates, R. D. (2017). The vulnerability and resilience of reef-building corals. *Current Biology* 27, R528–40. Available from: <http://dx.doi.org/10.1016/j.cub.2017.04.047>.

Reaka-Kudla, M. L. (1997) The global biodiversity of coral reefs: a comparison with rain forests. In M. L. Reaka-Kudla, D. E Wilson, and E. O. Wilson (eds) *Biodiversity II: Understanding and Protecting Our Biological Resources*. Washington, DC: Joseph Henry Press, pp. 83–108 Available from: <https://doi.org/10.17226/4901>.

Roche, R. C., Williams, G. J., and Turner, J. R. (2018) Towards developing a mechanistic understanding of coral reef resilience to thermal stress across multiple scales. *Current Climate Change Reports* 4 (1), 51–64. Available from: <https://doi.org/10.1007/s40641-018-0087-0>.

Romañach, S. S., DeAngelis, D. L., Koh, H. L., Li, Y., et al. (2018) Conservation and restoration of mangroves: global status, perspectives, and prognosis. *Ocean and Coastal Management* 154, 72–82. Available from: <https://doi.org/10.1016/j.ocecoaman.2018.01.009>.

Uthicke, S., et al. (2015) Outbreak of coral-eating Crown-of-Thorns creates continuous cloud of larvae over 320 km of the Great Barrier Reef. *Sci. Rep.* (Online) 5, 16885. Available from: doi: 10.1038/srep16885 (Accessed 14 January 2019).

Van Oppen, M. J. H., Gates, R. D., Blackall, L. L., et al. (2017) Shifting paradigms in restoration of the world's coral reefs. *Global Change Biology* (Online) 23, 3437–48. Available from: doi:10.111/gcb.13647. (Accessed 19 January 2019).

## Chapter 6: Deep-ocean biology

Anonymous (2018) The house that sank: creatures called giant larvaceans help ferry food—and pollution—to the deeps. *The Economist*, 22 and 28 September 2018, p. 68. Available from: <https://www.economist.com/science-and-technology/2018/09/20/giant-larvaceans-make-their-houses-from-mucus> (Accessed 16 January 2019).

Clark, M. R., Tittensor, D., Rogers, A. D. P., et al. (2006) *Seamounts, Deep-Sea Corals and Fisheries: Vulnerability of Deep-Sea Corals to Fishing on Seamounts beyond Areas of National Jurisdiction.* Cambridge: UNEP-WCMC.

Conniff, R. (2017) Up from the depths: the mass nighttime movement of life from deep sea up to surface is Earth's largest wildlife migration—a vertical feast that helps fuel the planet. (Online) Available from: <https://www.nwf.org/Magazines/National-Wildlife/2018/Dec-Jan/Animals/Vertical-Migration> (Accessed 24 January 2019).

Copley, J. (2014) Mapping the deep, and the real story behind the '95% unexplored' oceans. (Online) Available from: <http://moocs.southampton.ac.uk/oceans/2014/10/04/mapping-the-deep-and-the-real-story-behind-the-95-unexplored-oceans/.>.

Dubilier, N., Bergin, C., and Lott, C. (2008) Symbiotic diversity in marine animals: the art of harnessing chemosynthesis. *Nat Rev Micro* (Online) 6, 725–40. Available from: DOI:10.1038/nrmicro1992 (Accessed 19 January 2019).

Etnoyer, P. J. (2010) Deep-sea corals on seamounts. *Oceanography* 23 (1), 128–9.

The Five Deeps Expedition (n.d.) The world's first manned expedition to the deepest point in each of the five oceans. (Online) Available from: <https://fivedeeps.com/> (Accessed 6 July 2019).

Forest & Bird (n.d.) Best fish guide 2017. (Online) Available from: <http://bestfishguide.org.nz/> (Accessed 16 January 2019).

Katija, K., Choy, C. A., Sherlock, R. E., Sherman, A. D., and Robison, B. H. (2017) From the surface to the seafloor: how giant larvaceans transport microplastics into the deep sea. *Science Advances* (Online) 3 (8), e1700715 Available from: DOI: 10.1126/sciadv.1700715 (Accessed 16 January 2019).

Lampert, W. (1989) The adaptive significance of diel vertical migration of zooplankton. *Functional Ecology* 3, 21–7.

Marine Stewardship Council (2016) Orange Roughy: the extraordinary turnaround. From a troubled history to Marine Stewardship Council certification. (Online) Available from: <http://orange-roughy-stories.msc.org/> (Accessed 16 January 2019).

Nakagawa, S. and Takai, K. (2008) Deep-sea vent chemoautotrophs: diversity, biochemistry and ecological significance. *FEMS Microbiol. Ecol.* (Online) 65, 1–14. Available from: DOI:10.1111/j.1574-6941.2008.00502.x (Accessed 19 January 2019).

Ramirez-Llodra, E., Brandt, A. Danovaro, R. E., et al. (2010) Deep, diverse and definitely different: unique attributes of the world's largest ecosystem. *Biogeosciences Discussions*, 7, 2361–485.

Roark, E. B., Guilderson, T. P., Dunbar, R. B., Fallon, S. J., and Mucciarone, D. A. (2009) Extreme longevity in proteinaceous deep-sea corals. *PNAS* (Online)106 (13), 5204–8. Available from: <https://doi.org/10.1073/pnas.0810875106> (Accessed 16 January 2019).

Smith. C. (2012) Chemosynthesis in the deep-sea: life without the sun. *Biogeosciences Discussions* (Online) 9, 17037–52. Available from: doi:10.5194/bgd-9-17037-2012 (Accessed 15 January 2019).

Smith, C. R. and Baco, A. R. (2003) Ecology of whale falls at the deep-sea floor. *Oceanography and Marine Biology: An Annual Review*, 41, 311–54.

Wikipedia (2019) Challenger Deep. (Online) Available from: <https://en.wikipedia.org/wiki/Challenger_Deep> (Accessed 14 January 2019).

Woods Hole Oceanographic Institution (2015) Making organic molecules in hydrothermal vents in the absence of life. (Online) Available from: <https://www.whoi.edu/news-release/methane-formation> (Accessed 16 January 2019).

WoRDSS (n.d.) World Register of Deep-Sea Species. (Online) Available from: <http://www.marinespecies.org/deepsea/> (Accessed 14 January 2019).

## Chapter 7: Intertidal life

Anonymous (2003) *Oil in the Sea III: Inputs, Fates, and Effects*. Washington, DC: National Academies Press. Available from: <http://www.nap.edu/catalog/10388.html>.

Castilla, J. C. and Duran, L. R. (1985) Human exclusion from the rocky intertidal zone of central Chile: the effects on *Concholepas concholepas (Gastropoda)*. *Oikos* 45, 391–9.

Connell, J. H. (1961) The influence of intra-specific competition and other factors on the distribution of the barnacle *Chthamalus stellatus*. *Ecology* 42, 710–23.

Michel, J., Esler, D., and Nixon, Z. (2016) *Studies on Exxon Valdez Lingering Oil: Review and Update on Recent Findings—February 2016*. Exxon Valdez Oil Spill Trustee Council.

Paine, R. T. (1994) Marine rocky shores and community ecology: an experimentalist's perspective. Ecology Institute, Oldendorf/Luhe, Germany. Available from: <https://www.int-res.com/articles/eebooks/eebook04.pdf>.

Smith, J., Fong, P., and Ambrose, R. (2008) The impacts of human visitation on mussel bed communities along the California coast: are regulatory marine reserves effective in protecting these communities? *Environmental Management* 41 (4), 599–612.

Tomanek, L. and Helmuth, B. (2002) Physiological ecology of rocky intertidal organisms: a synergy of concepts. *Integrative and Comparative Biology* 42 (4), 771–5. Available from: <https://doi.org/10.1093/icb/42.4.771>.

## Chapter 8: Food from the oceans

FAO (2018) *The State of World Fisheries and Aquaculture 2018—Meeting the Sustainable Development Goals*. Rome. Licence: CC BY-NC-SA 3.0 IGO.

Gaines, S. D., Costello, C., Owashi, B., et al. (2018) Improved fisheries management could offset many negative effects of climate change. *Sci. Adv.* (Online) 4 (8), eaao1378. Available from: DOI: 10.1126/sciadv.aao1378. (Accessed 20 January 2019).

Gill, D. A., Mascia, M. B., Ahmadia, G. N., et al. (2017) Capacity shortfalls hinder the performance of marine protected areas globally. *Nature* 543, 665–9.

IFFO. The Marine Ingredients Organisation (2015) Fish In: Fish Out (FIFO) ratios for the conversion of wild feed to farmed fish, including salmon. (Online) Available from: <http://www.iffo.net/fish-fish-out-fifo-ratios-conversion-wild-feed> (Accessed 17 January 2019).

Lotze, H. K., Coll, M., and Dunne, J. A. (2011) Historical changes in marine resources, food-web structure and ecosystem functioning in the Adriatic Sea, Mediterranean. *Ecosystems* 14, 198–222.

Marine Stewardship Council (n.d.) Enjoy the seafood you love. (Online) Available from: <https://www.msc.org/> (Accessed 17 January 2019).

Montañez, A. (2018) How much of the world's protected land is actually protected? Available from: <https://www.scientificamerican.com/article/how-much-of-the-worlds-protected-land-is-actually-protected1/> (Accessed 17 January 2019).

Pauly, D. and Zeller, D. (2016) Catch reconstructions reveal that global marine fisheries catches are higher than reported and declining. *Nat. Commun.* (Online) 7, 10244. Available from: doi: 10.1038/ncomms10244 (Accessed 17 January 2019).

Roberts, C. M., O'Leary, B. C., McCauley, D. J., et al. (2017) Marine reserves can mitigate and promote adaptation to climate change. *PNAS* (Online) 114 (24), 6167–75. Available from: <http://www.pnas.org/cgi/doi/10.1073/pnas.1701262114> (Accessed 20 January 2019).

Ryther, J. H. (1969) Photosynthesis and fish production in the sea. *Science* 166, October, 72–6.

Steneck, R. S. (2012) Apex predators and trophic cascades. *PNAS* (Online) 109 (21), 7953–4; DOI:10.1073/pnas.1205591109 Available from: <https://doi.org/10.1073/pnas.1205591109> (Accessed 17 January 2019).

Travis, J., Coleman, F. C., Auster, P. J., et al. (2014) Integrating the invisible fabric of nature into fisheries management. *PNAS* (Online) 111 (2), 581–4. Available from: <https://doi.org/10.1073/pnas.1305853111> (Accessed 17 January 2019).

UNEP-WCMC and IUCN (2019) Marine protected planet. (Online) Available from: <http://www.protectedplanet.net. (Accessed 17 January 2019).

Watson, R. A. and Tidd, A. N. (2018). Mapping nearly a century and a half of global marine fishing: 1869 to 2015. *Marine Policy* 93, 171–7. Available from: <https://doi.org/10.1016/j.marpol.2018.04.023>.

WWF (2017) Get to know your seafood. (Online) Available from: <http://wwf.panda.org/get_involved/live_green/out_shopping/seafood_guides/> (Accessed 17 January 2019).

Ye, Y., Cochrane, K., Bianchi, G., et al. (2013) Rebuilding global fisheries: the World Summit goal, costs and benefits. *Fish and Fisheries* 14, 174–85. DOI: 10.1111/j.1467-2979.2012.00460.x.

# Chapter 9: The future of our oceans

Grip, K. (2017) International marine environmental governance: a review. *Ambio* (Online) 46, 413–27. Available from: DOI 10.1007/s13280-016-0847-9. (Accessed 20 January 2019).

Jackson, J. B. C. (2008) Ecological extinction and evolution in the brave new ocean. *Proceedings of the National Academy of Science of the USA*, 105 Suppl. 1, 11458–65.

Jackson, J. B. C. (2010) The future of the oceans past. *Philosophical Transactions of the Royal Society of London* B 365, 3765–8.

Jones, K. R., Klein, C. J., Halpern, B. S., et al. (2018) The location and protection status of Earth's diminishing marine wilderness. *Current Biology* (Online) 28, 1–7. Available from: <https://doi.org/10.1016/j.cub.2018.06.010>. (Accessed 20 January 2019).

Lotze, H. K. and McClenachan, L. (2013) Marine historical ecology: informing the future by learning from the past. In M. D. Bertness, J. F. Bruno, B. R. Silliman, and J. J. Stachowicz (eds) *Marine Community Ecology and Conservation*. Sunderland, Mass.: Sinauer. Available from: <https://www.lenfestocean.org/-/media/legacy/lenfest/pdfs/mcec_ch08.pdf?la=en&hash=41C7D8770BC6AB33744FB13D40CD5A3DB111EDF0>.

McCauley, D. J., Pinsky, M. L., Palumbi, S. R., et al. (2015) Marine defaunation: animal loss in the global ocean. *Science* (Online) 347 (6219), 1255641. Available from: DOI: 10.1126/science.1255641. (Accessed 20 January 2019).

Rockström, J., Gaffney, O., Rogelj, J., et al. (2017) A roadmap for rapid decarbonization. *Science* (Online) 355 (6331), 1269–71. Available from: doi: 10.1126/science.aah3443. (Accessed 20 January 2019).

Visbeck, M. (2018) Ocean science research is key for a sustainable future. *Nature Communications* (Online) 9 (690), 1–4. Available from: DOI: 10.1038/s41467-018-03158-3. (Accessed 20 January 2019).

# Index

For the benefit of digital users, indexed terms that span two pages (e.g., 52–53) may, on occasion, appear on only one of those pages.

Marine Biology

# DESERTS
## A Very Short Introduction
### Nick Middleton

Deserts make up a third of the planet's land surface, but if you picture a desert, what comes to mind? A wasteland? A drought? A place devoid of all life forms? Deserts are remarkable places. Typified by drought and extremes of temperature, they can be harsh and hostile; but many deserts are also spectacularly beautiful, and on occasion teem with life. Nick Middleton explores how each desert is unique: through fantastic life forms, extraordinary scenery, and ingenious human adaptations. He demonstrates a desert's immense natural beauty, its rich biodiversity, and uncovers a long history of successful human occupation. This *Very Short Introduction* tells you everything you ever wanted to know about these extraordinary places and captures their importance in the working of our planet.

# AMERICAN POLITICAL PARTIES AND ELECTIONS

## A Very Short Introduction

Sandy L. Maisel

Few Americans and even fewer citizens of other nations understand the electoral process in the United States. Still fewer understand the role played by political parties in the electoral process or the ironies within the system. Participation in elections in the United States is much lower than in the vast majority of mature democracies. Perhaps this is because of the lack of competition in a country where only two parties have a true chance of winning, despite the fact that a large number of citizens claim allegiance to neither and think badly of both. Studying these factors, you begin to get a very clear picture indeed of the problems that underlay this much trumpeted electoral system.

# CANCER
## A Very Short Introduction
### Nick James

Cancer research is a major economic activity. There are constant improvements in treatment techniques that result in better cure rates and increased quality and quantity of life for those with the disease, yet stories of breakthroughs in a cure for cancer are often in the media. In this *Very Short Introduction* Nick James, founder of the CancerHelp UK website, examines the trends in diagnosis and treatment of the disease, as well as its economic consequences. Asking what cancer is and what causes it, he considers issues surrounding expensive drug development, what can be done to reduce the risk of developing cancer, and the use of complementary and alternative therapies.

# SCIENTIFIC REVOLUTION
## A Very Short Introduction
Lawrence M. Principe

In this *Very Short Introduction* Lawrence M. Principe explores the exciting developments in the sciences of the stars (astronomy, astrology, and cosmology), the sciences of earth (geography, geology, hydraulics, pneumatics), the sciences of matter and motion (alchemy, chemistry, kinematics, physics), the sciences of life (medicine, anatomy, biology, zoology), and much more. The story is told from the perspective of the historical characters themselves, emphasizing their background, context, reasoning, and motivations, and dispelling well-worn myths about the history of science.

www.oup.com/vsi